STEPHEN HAWKING: Quest for a Theory of Everything

'Ferguson deftly conveys an exciting sense of both Stephen Hawking himself and his preeminent role in modern-day theoretical physics . . . A fine introduction to Hawking's own *A Brief History of Time*' **Booklist**

'Fascinating . . . our taste for wonder and a certain morbidity is well catered for' **Anthony Burgess,** *Observer*

'Kitty Ferguson has taken a probing, sympathetic look at the lives behind the fame . . . The author deals delicately with the human side of the Hawking phenomenon' *Sunday Times*, **Perth**

'The reader is drawn on by some of the best scientific definitions, analogies, simple explanations and downright interesting writing I have ever seen' *London Evening Standard*

'An explanation of what the bloody hell *A Brief History of Time* was all about, for those who didn't get as far as chapter two' *Venue*

'Kitty Ferguson is a brilliant intermediary between the thinking of the physicist and the thinking of ordinary people' *German National Radio*

'Splendid, carefully researched study of the man and his ideas' *Kirkus Reviews*

Also by Kitty Ferguson

BLACK HOLES IN SPACETIME

STEPHEN HAWKING: QUEST FOR A THEORY OF EVERYTHING

THE
FIRE IN THE
EQUATIONS

Kitty Ferguson

BANTAM BOOKS
TORONTO · NEW YORK · LONDON · SYDNEY · AUCKLAND

THE FIRE IN THE EQUATIONS
A BANTAM BOOK : 0 553 40593 4

Originally published in Great Britain by Bantam Press,
a division of Transworld Publishers Ltd

PRINTING HISTORY
Bantam Press edition published 1994
Bantam edition published 1995

Set in Linotron Times by
County Typesetters, Margate, Kent.

Bantam Books are published by Transworld Publishers Ltd,
61–63 Uxbridge Road, Ealing, London W5 5SA,
in Australia by Transworld Publishers (Australia) Pty Ltd,
15–25 Helles Avenue, Moorebank, NSW 2170,
and in New Zealand by Transworld Publishers (NZ) Ltd,
3 William Pickering Drive, Albany, Auckland.

Printed and bound in Great Britain by
Cox & Wyman Ltd, Reading, Berks.

*To my father and mother,
Herman and Tina Vetter*

A WORD ABOUT INCLUSIVE LANGUAGE

The author of a book on the topic of science and religion needs a pronoun for God. Regardless of whether I choose to call God 'he' or 'she', I find myself making a statement which I don't wish to make. Using them interchangeably seems contrived and gets confusing. 'She/he' or 'he/she' is cumbersome . . . and one still has the problem of which gender comes first in the pairing. 'It' will not do. Lacking a better solution, I have chosen to use 'he', which makes the weaker statement and is more easily interpreted as inclusive.

ACKNOWLEDGEMENTS

The author wishes to thank the following, who read all or part of this manuscript and answered her questions, for their valuable suggestions, comments and corrections:

Richard Dawkins, Savas Dimopoulos, Andrew Dunn,
Caitlin Ferguson, Yale Ferguson, Joseph Ford,
Matthew Fremont, Alan Guth, Marissa Herrera,
Stephen Hawking, David Hegg, P. Susie Maloney,
Robert Maseroni, D. Paul La Montagne, Sir Brian Pippard,
John Polkinghorne, Herman Vetter, Tina Vetter,
Steven Weinberg.

Also the following for sending me interesting clippings,
letters, and other material, for their ideas and suggestions,
and for their assistance in many other ways:

Serafina Clarke, Chuck Collins, Colin Ferguson, Duff Ferguson,
Howard Ferguson, Peter Harmon, Howard Helms,
Richard Langhorne, Jim Morgan, Peter Ochs, Allan Sandage,
Veronica Towle, Richard Westfall.

ACKNOWLEDGMENT

CONTENTS

1 'They Buried Him in Westminster Abbey'   1

2 Seeing Things   4
   Is the rational universe an illusion?   12
   'In Nature's infinite book of mysteries . . .' can
     we read very much at all?   19
   Is objective reality a mirage?   25
   Are we really free agents?   30
   Is the universe a uni-verse?   33

3 Almost Objective   35
   Where is fancy bred?   37
   The spectacles-behind-the-eyes   44
   The muse of science: Is truth beautiful?   59
   Does truth surpass proof?   63
   The elite of science   66
   The spirit of the times   69
   The essential Godlessness of science   73
   At the limits of scientific truth   78
   First steps beyond the mind's-eye view   80
   Is there anything else?   81
   The insidiousness of God   86
   The morality of science: Is truth good?   87

4 Romancing the Creation   89
   The uncomfortable concept of a beginning   90
   The Gordian knot of singularity   102

The magic of imaginary time 108
The pulsing universe and the elusive clue of
dark matter 117
The mysterious wobbling of nothingness 123
'Reality (whatever that may be)' 126
Reality in the absence of apples 129
What place for a creator? 134
The third candidate 137
The mother of all chicken-and-egg stories 139

5 The Elusive Mind of God 143
God as the embodiment of the laws of physics 145
A presence behind the process 146
The leap to purpose: The God who wishes to
drink tea 147
The watchmaker 149
The universe as a 'put-up job' 163
Second Gordian knot: The anthropic principle 164
Hacking at the second Gordian knot 166
The inflationary universe 167
Baby universes to the rescue! 171
Not the ether again! 173
The longing of Johannes Kepler 178
The fiddler on the roof 184

6 The God of Abraham and Jesus 185
The law-breaker 189
The hard edge of legalism 191
The soft underbelly of legalism 195
The death of the God of the Gaps 204
Chaos meets Control 205
'Top-down' determinism? 221
'I AM' 225
When truths collide 228
The ultimate self-confirming hypothesis 231

The masterful use of parallel perfect fifths   234
Who is the 'I' in 'I AM'?   239

7   Inadmissible Evidence   241
    Public vs. private knowledge   242
    Admissible evidence?   245
    The spectacles-behind-the-eyes, revisited   246
    The cloud of witnesses   247
    A game of 'I Doubt It'   251
    The Lucy problem   253
    'I should not believe such a story were it told
      me by Cato!'   254
    'The Invincible Ignorance of Science'   259
    'For the Bible tells me so' – the evidence of
      scripture   260
    Is there proof in the pudding? The evidence of
      results   263
    Armchair truth: The argument from reason   266
    The argument from explanatory power   270
    The argument from nature   274
    The argument from availability   277

8   Theory of Everything . . . Mind of God   279

   Notes   285

   Bibliography   295

   Index   303

1

'THEY BURIED HIM IN WESTMINSTER ABBEY'

AT 8 O'CLOCK IN THE EVENING OF TUESDAY, 25 APRIL 1882, THE horse-drawn funeral car carrying Charles Darwin's coffin arrived at Westminster Abbey. The sixteen-mile journey in the rain from the Kentish village of Downe had taken all day. The coffin was borne through the cloisters of the Abbey and placed in the Chapel of St Faith, a spare, sepulchral, vaulted chamber, ice-cold and lit only by two flickering lanterns. It was a magnificent coffin, but not the coffin he and his family had wanted. That had been an oak box, 'all rough, just as it left the bench, no polish, no nothin',' said John Lewis, the Downe village carpenter who built it. 'When they agreed to send him to Westminster . . . my coffin wasn't wanted. This other one you could see to shave in.'[1] But Charles Darwin belonged to the nation now and to history, not to his family and his village, and at noon the following day he would be buried in state in the Abbey.

On the previous Sunday the news of Darwin's death had brought forth paeans of praise for him and his scientific discoveries from the pulpits of London, and the newspapers had continued the theme: 'Darwin's doctrine is in no wise inconsistent with strong religious faith and hope,' proclaimed the *Daily News*.[2] 'True Christians can accept the main scientific facts of Evolution just as they do of Astronomy and Geology, without any prejudice to more ancient and cherished beliefs,' pontificated the *Standard*.[3] Canon H. P. Liddon, in an afternoon sermon in St Paul's Cathedral, compared Darwin to St Thomas – 'doubting'

Thomas. Canon Liddon chose not to condemn Darwin's religious scepticism but to commend 'the patience and care with which he observed and registered minute single facts'. St Thomas had refused to believe in Christ's resurrection unless he could put his hand into the wounds inflicted during the crucifixion. Darwin, like Thomas, had insisted on evidence, what Canon Liddon called 'the clearly ascertained report of the senses'.[4] The *Guardian* reassured its readers that they should not have 'any misgivings lest the sacred pavement of the Abbey should cover a secret enemy of the Faith'. The honour of burial there should be seen as 'a happy trophy of the reconciliation between Faith and Science'.[5]

What? Hadn't Darwin ended any possibility of believing strongly in both science and the Judaeo-Christian God without indulging in intellectual dishonesty? Extremes of opinion among both scientists and religious people ever since would certainly have it so. Darwin demolished the literal interpretation of the biblical Creation story and undermined one of the most eloquent arguments for the existence of God, that the world was a place perfectly designed for the survival and sustenance of human beings. Evolution and survival of the fittest provided a natural explanation for what had seemed a miracle. Yet there have been many scientists since Darwin, and there are many even now in the 1990s, who are devout believers in God. Do they, as someone said of physicist Max Planck, forget their faith when they go into the lab, and forget their science when they go into church?

On 26 April 1882, the skies were still leaden. The gas-lit Abbey was dank and gloomy, thronged with sombrely dressed luminaries of government and science as well as middle-class citizens who came without black-bordered tickets and were allowed to fill the less desirable seats. The funeral was a religious service with readings and anthem texts from the Gospels and the Psalms. The Abbey organist, J. Frederick Bridge, had composed an anthem to be sung for the occasion. He had chosen words from the Book of Proverbs: 'Happy is the man that findeth wisdom, and getteth understanding.'[6] Later the chief mourners and the public filed past the grave to the accompaniment of the 'Dead March' from Handel's *Saul,* a march which in the original was a dirge for a king who had torn himself away from the love of God to rely on the power of himself.

What did they make of it, the mourners, the dignitaries, and the merely curious at the funeral of Charles Darwin? Is the Wisdom of science the Wisdom of Proverbs? Proverbs also describes a single-minded human struggle which ends in the gift of 'the knowledge of God'.[7] A century after Darwin's death, another great English scientist, Stephen Hawking, wrote that the ultimate triumph of human reason would be to know the Mind of God. He said science could get us almost there, but not the whole way. Is the Knowledge of God in Proverbs the Mind of God in *A Brief History of Time*? Or is Hawking's a metaphor for our becoming God-like in our complete knowledge? Is there a Person waiting for us at the end of the quest, or is that Person *us*, reasoning humanity triumphant, evolution's masterpiece?

Ultimate reality, whatever that turns out to be, is the end of the quest. Paradoxically, it must also be the beginning. We must ask whether there is anything about our universe, about ourselves, that we can take for granted – any fundamental we can use as a starting place for the exploration of everything else. If it is difficult to find such a 'still point' – and we shall find that it is indeed difficult – then the quest for ultimate truth must begin with a leap of faith. Not faith that we are capable of complete understanding. Faith that we can know anything at all.

2

SEEING THINGS

Kick at the rock, Sam Johnson, break your bones,
But cloudy, cloudy is the stuff of stones.

FROM *EPISTEMOLOGY* BY RICHARD WILBUR

THERE IS AN OLD STRAIGHT-BACKED OAK CHAIR STANDING AGAINST
the wall across from my desk. It was made by hand about a
century ago in the Texas hills, when that hill country was still a
frontier. I inherited the chair from my grandparents. When my
grandmother and grandfather looked at it in the dining room of
their Mason County parsonage, they saw the same chair I see
here today in my study, or so I assume. Maybe the wood has
darkened a little with age. Someone visiting me today will see the
same chair my grandparents saw and that I see, or so I assume.
Common sense tells me I'm right.

My faith in common sense is a faltering faith. In writing my
previous book, *Stephen Hawking: Quest for a Theory of
Everything,* I explored a world that was not on any level a
common-sense world. A man of extraordinary genius condemned
to live out his years locked in a useless body without movement or
speech, whose sheer bloody-minded courage nevertheless allows
him to be one of the pre-eminent physicists of the century as well
as an international celebrity – Hawking is not a common-sense
figure. Quantum mechanics and Einstein's general relativity are
not common-sense subjects. Nevertheless, having made that
journey through the looking glass and back, having seen for

4

myself how absurd and counter-intuitive the world is, I still sit here and say, Yes, the reality of that oak chair was the same for my grandparents as it is for me today.

I recently re-read Sir Arthur Eddington's introduction to his book *The Nature of the Physical World*,[1] in which he speaks of a table as I am writing about my chair. There is a story about Eddington that when someone remarked that only three people in the world understood Einstein's theory of relativity, he muttered, 'I'm trying to think who the third could be.' But Eddington, for all his remarkable intellect, also had a talent for taking complicated scientific concepts and explaining them in simple English. In the paper I've been reading he describes a piece of furniture like my chair as seen through the eyes of physics. It is not a description my grandparents would have recognized.

My chair is made up of atoms, and atoms are almost entirely empty space. That means my chair consists in very large part of emptiness. My chair is a blur of uncertainty, which I'm allowed to think of as unimaginably tiny particles whizzing around in a fuzzy manner. I know I mustn't think of these particles as 'things' in exactly the sense I think of the chair as a 'thing' – something that can be pinned down in the accurate way we expect to pin 'things' down. I wonder whether a chair consisting of 'non-things' can itself fairly be called a 'thing', and why I see it as such. Is my familiar chair more real than the same chair as Eddington describes it? Or must I consider the smallest level of the universe the most 'real'? We shall get back to those questions later. My chair looks real enough to me.

A perfectly common-sense, familiar Texas oak chair. That seems to be the only interpretation anyone's five senses can make of it. If I touch the chair seat, a swarm of electronic impulses bats against my hand, which is also a swarm of electronic impulses. The combined bulk of these impulses is less than a billionth of the bulk of the chair itself. Thus all that empty space. But somewhere between the electronic impulses and my consciousness a mysterious transformation occurs which causes me without any effort to interpret all of this as a solid piece of oak.

Perhaps that interpretation is the only possible interpretation on our level of the universe, but I am curious as to whether you really would see and feel the same chair if you were here in my study. We would describe it to one another in more or less the same way, but our descriptions would have to consist of words and would have to depend entirely upon the mental images each of us has learned to associate with those words. Have you perhaps learned to associate the word 'brown' in your mind's eye with a different hue from that which my mind conjures up when 'brown' is mentioned? My chair in my mind's eye is surely not precisely my chair in your mind's eye.

What do you and I really know about chairs or anything else? How do we know it? We humans have gone a long way beyond such modest observations of the world around us. We trust not only our five senses but a wealth of accumulated findings and a spectacularly complex system of mathematics and logic. From all of this we hope to find out the truth about far, far more than chairs and tables. What is the universe? How did it begin? What happened before that? How and when will it end? What is space and, even more puzzling, what is time? We hope to be able to answer Hawking's question 'Why does the universe go to all the bother of existing?'[2] We hope, with him, to know the mind of God.

We also would like to know the answers to questions left unspoken in Hawking's *A Brief History of Time,* questions which nevertheless cry out from its pages for answers. Why? Why should a man be dealt such a preposterously unbalanced fate – appalling disease, extraordinary genius, bloody-minded courage? It isn't just Stephen Hawking's dilemma. In a sense, the cynic might suggest, it sums up the situation of the entire human race. It is the human condition, a mockery of rationality, a theatre of the absurd.

My grandparents would have bowed to all of it as the work of a God whose activities are far beyond our understanding, a God whose 'tough love' goes beyond that of any human parent. That is how they dealt with the absurdity of their youngest son being blown up with his boat in the English Channel, not by enemy fire but by a stray American shell. Some of the rest of us aren't willing to take that sort of explanation lying down, and neither, really, were my grandparents – not without complaint and some rebellion and more than a few demands for clearer explanations, demands directed to a God they were quite sure could give the explanations if he chose. The mind of God to my grandparents

6

was not something to be learned through physics, though my grandfather was avidly interested in whatever scientists could tell him, and certainly didn't think such information irrelevant to his own personal spiritual quest.

Hawking does not share my grandparents' faith in God. But he shares their curiosity about ultimate truth, ultimate explanation. Like them he longs to have all illusions swept away, to know the unveiled truth behind everything, no holds barred. To act on such longing involves great risk. Does the atheist want to know the truth if the truth is that there is a God? Does the believer want to know the truth if the truth is that there is not? Are we that open-minded?

There is a further element of risk for anyone on a search for truth. You cannot start in a vacuum. You must begin by trusting some ideas about the universe that have never been proved, may never be proved, and might turn out to be wrong. To be simplistic about it, you have to assume that you exist and that you are sane. Those may not be such difficult assumptions. Common sense supports them. Of course you have to believe they are true in order to trust your common sense. You see what sort of mental maze we get ourselves into!

The search for truth in science is based on agreement concerning just such basic assumptions. It is a gamble, if you will; a gamble that certain articles of faith which cannot be proved by science are nevertheless well-founded enough to provide a springboard for all scientific investigation. It is intriguing to find that religion shares much of science's basic view of reality. How is it that two approaches, science and religion, both claiming to be avenues to truth but in many ways reputed to clash with one another, should be in agreement on so basic a level?

The explanation could be quite simple – that we are all looking at the same universe, and what is obvious to one reasonable person is equally obvious to the next. If that is so, it should not surprise us to find all reasonable people more or less in agreement about certain fundamental aspects of the universe. However, the agreement is not unanimous. We are speaking of a world-view shared by science (since the seventeenth century) and Western religion, with exceptions even here, but not shared by all of humanity who presumably experience and have experienced the same universe.

7

Perhaps the explanation lies in the origins of science as we know it today. Scientists of the seventeenth century, most but not all of whom had religious views closer to my grandparents than to Hawking (many of the first Fellows of the Royal Society in England were Puritans), developed a procedure to be used in the search for scientific knowledge, a procedure that would systematically separate what is true from what is not true. That is the procedure we call the scientific method. It has served us splendidly ever since its birth and made our spectacular technology possible. Whatever the scientific method's origins or its philosophical foundations, we have no cause to doubt its usefulness.

Depending upon whether we believe in God, you or I might leave God out of the following articles of faith, but otherwise we would find little in this seventeenth-century world-view with which to disagree. In the seventeenth century a scientist could have had it both ways without risking charges of contradiction. What he learned from his religion and his direct experience of the universe led him to believe the following:

- The universe is *rational*, reflecting both the intellect and the faithfulness of its Creator. It has pattern, symmetry, and predictability to it. Effect follows cause in a dependable manner. For these reasons, it is not futile to try to study the universe.

- The universe is *accessible* to us, not a closed book but one open to our investigation. Minds created in the image of the mind of God can understand the universe God created.

- The universe has *contingency* to it, meaning that things could have been different from the way we find them, and chance and/or choice have played a role in making them what they are. Whether this is contingency in the sense that chance and choice play an on-going role within the universe, or merely in the sense that there was an initial chance occurrence or choice which brought about this universe instead of a different one or none at all, one cannot learn about the universe by pure thought and logic alone. Knowledge comes by observing and testing it.

- There is such a thing as *objective* reality. Because God exists and sees and knows everything, there is a truth behind everything. Reality has a hard edge to it and does not cave in or

8

shift like sands in the desert in response to our opinions, perceptions, preferences, beliefs, or anything else. Reality is not a democracy. There is something definite, some raw material, out there for us to study.

- There is *unity* to the universe. There is an explanation – one God, one equation, or one system of logic – which is fundamental to everything. The universe operates by underlying laws which do not change in an arbitrary fashion from place to place, from minute to minute, or even millennium to millennium. There are no loose ends, no real contradictions. At some deep level, everything fits.

Divorced now from the assumption that there is a God, these five assumptions about the universe, these five articles of faith, if you will – rationality, accessibility, contingency, objectivity, and unity – continue to underlie the practice of science. Some would argue that upon them depends all possibility of doing science as we know it. The best argument for their validity is not that they are obvious but that the scientific method seems to work so well! The proof (dangerous word) is in the pudding.

Nevertheless, we are left with some questions. Is the scientific method, which serves us so admirably in our quest for knowledge about the physical universe, also a reliable source of complete understanding about the events around us and of our own existence? If the scientific method and the approach of constructing mathematical models cannot answer Hawking's question 'Why does the universe go to all the bother of existing?', what can? Is there a meaning and is there a God (or 'mind of God') beyond the reaches of the scientific method but not beyond the reaches of human reason?

Human reason cannot be divorced from common sense, which says: I can see that the universe has rationality and accessibility and contingency and objectivity, and so can most of the people I know . . . If other cultures look at the same universe and draw different conclusions, well, that's certainly mysterious but I can't be too much bothered by it . . . Maybe they're wrong . . . I have to trust my senses.

If you feel that way you may be accused of being naive. However, some very un-naive people would back you up to the extent of saying that the argument 'This is what I make of it all, and I don't have any stronger reference point than that' is in fact very hard to refute. Sir Brian Pippard of Cambridge University, the physicist who introduced me to the Eddington book I mentioned earlier, tells me there are more chairs across from me in my study than just the common-sense chair and the chair-as-seen-by-physics. We've already mentioned a third, but we didn't give it quite the importance Brian Pippard wants us to. It is the chair in my mind's eye, an image I can't share with Pippard or you or anybody, because I can't let you into my mind to see whether 'brown' or anything else looks the same in my mind as in yours. We can discuss my chair, even compare it with a description my grandfather wrote in an inventory of his furniture, and come easily to the conclusion that we are all talking about the same object, but our mind's-eye chairs will not all be identical. Our interpretations of whatever it is out there across from my desk will not be exactly the same.

Perhaps the mind's-eye chair seems to you less substantial, representing a fuzzier and more subjective viewpoint, than Eddington's chair-as-viewed-by-physics or the common-sense chair we thought was there before we began all this talk about it. Evidence coming from one person is not so dependable as something you and I and others could agree upon precisely. The scientific method cannot accept such individual, uncorroborated evidence. But Pippard argues, and it is hard to take exception with him, that the *one and only* certainty each of us has is the certainty of his or her own existence. What this means is that 'Come what may, it is the [chair] in the mind of each of us to which all else we believe in must conform.'[3] Of course even the certainty of my own existence is questionable. Philosophically it is possible to argue that I do not exist. But I notice that I do, and that is the only reference point I have to go on. I am, by default, my unique authority in the matter. I also have only my presumption of my existence and my mind's-eye images to go on if I want to come to any conclusions at all about my chair or the rest of the universe.

What has happened to objective truth if truth in my mind's eye may be different from truth in yours? Pippard isn't saying that what my mind's eye leads me to believe is truth. What your mind's eye leads you to believe isn't necessarily truth either. Pippard is saying that the one and only certainty *I* have is of *my* own existence. The only certainty *you* have is of *your* own existence. Each of us has only that as a starting point. The question is, how does what begins with my certainty of my existence, and continues with my mind's-eye view of the universe, end with the discovery of objective truth – even perhaps with that ultimate distillation of objective truth, the Theory of Everything or the Mind of God? What makes me think I can begin HERE and arrive THERE where ultimate, objective truth is in my mind's-eye view?

One of the articles of faith listed above was that truth does exist in a way that is independent and 'other' from myself or yourself, unchanged whether or not it is studied by a physicist or a common-sense observer and not affected by how it is viewed in anyone's mind's eye. Pippard tells me there is a fourth chair across the room from me – the 'chair-as-it-is-in-itself', the most bed-rock solid of all views of my chair and the most elusive.

I *would* like to know whether my perception of my grandparents' chair and the rest of the universe has any relation to ultimate reality. If there is a God, I would like to know what it all looks like from God's vantage point. Sir Brian Pippard says my chair-as-it-is-in-itself – and, by extension, the universe-as-it-is-in-itself – might turn out to be 'something quite other, outside the range of our thought.'[4] How much more 'other' might be the Mind of God?

To bring us down out of the clouds to a more practical level, suppose you decide, on a quest for knowledge, to attach particular importance to what scientists have discovered about the universe by means of the scientific method, which does after all seem to be a very reliable method for finding out what is what. If you proceed along these lines you may be in for a shock. You will learn not only that science has not proved the assumptions that the universe is rational, accessible, contingent, objective, and has unity to it, but also that there have been scientific discoveries and theories which lead us to question seriously whether those assumptions are correct. Where does that leave us? Are the foundations of all our knowledge crumbling? Is the search for truth about to self-destruct? Can we know ANYTHING?

11

IS THE RATIONAL UNIVERSE AN ILLUSION?

We speak of a universe that is rational and logical, a universe that makes sense and has pattern to it. Strong evidence of this rationality is the dependability of cause and effect. Everyone knows that nothing happens without something causing it to happen. The cause may be obvious or it may be hidden beyond our ability ever to discover what it is, but it is always there, or so we assume. We conduct lengthy and expensive investigations to find the cause of a disaster like the explosion of the United States space-shuttle *Challenger*. Extremely cold weather, a problem with the O-ring seals. No-one thinks seriously of concluding 'It just happened, nothing more to be said about it.' Every effect has a cause, and that means there are chains of cause and effect, chains which we don't expect to come to a dead end.

Even if chance played a part – the perhaps unlikely instance of weather conditions and O-ring problem coinciding – no-one would claim that the 'law' of cause and effect had been broken. The weather conditions had a history of cause and effect behind them and so did the O-rings and the adhesive that secured them. Too complicated to follow, perhaps, but still there. A story was involved, and if we could find out what the story was, we could explain the disaster. If we had failed to find some link in the story, we still wouldn't have leapt to the conclusion that no such link existed. It would not occur to us to do that. We say the evidence is insufficient.

We've grown accustomed to the way cause and effect operate on our level and in the part of the universe we can observe, and so it seems safe to assume, though we have no way of demonstrating it beyond a certain point, that cause and effect similarly operate in areas of the universe that we cannot observe directly. We rely on this being so. We think that cause and effect will continue to operate in the future, with no real guarantee that today isn't the last day they will be in operation. If an experiment gives one result today, it ought to give the same result tomorrow. If it fails to do so, we question the experiment or our interpretation of it, not the reliability of the concept of cause and effect.

We also assume with no way of testing it that cause and effect operated at the very earliest stages of the universe. Even at the moment of creation? So strong is our belief that it is difficult to

12

imagine that the universe itself could exist without a cause; that it could just BE. We want to know how it happened, and we want to find the answer to the question 'Why?' Or even 'Who?'

Belief in cause and effect is a cornerstone of the scientific method. Nevertheless some scientists keep reminding us that the 'law' of cause and effect is an 'article of faith', not a law at all. It can't be proved to operate in all cases. Indeed, there is a major subfield of modern physics that requires us to reconsider our assumption that every event has an unbroken history of cause and effect leading up to it.

'Quantum mechanics' is not a name like 'black hole' or 'quasar' to light the fires of our imagination. Yet the study of the quantum level of the universe is an area of physics which seems as exotic as anything ever dreamed of in science fiction. It is the study of the smallest size levels in the universe, of atoms and elementary particles. Some of what happens on that level is extremely difficult to explain in a way that satisfies our wish for a common-sense description. One of the oddities is that we observe individual events which are, in a sense, 'uncaused' events, happenings without a history of the sort we normally assume any event must have.

The quantum level of the universe will crop up repeatedly in this book. For the benefit of those who haven't already learned something about it, or have forgotten what they used to know, let us pay it a preliminary visit before proceeding further:

Picture something relatively familiar, our solar system. The planets orbit the sun in orbits that we have learned to predict. At any given moment each planet in relation to the others has a definite position and is travelling in a definite direction at a definite speed. We can see that Saturn is THERE today, and, because we also know its speed and direction of movement, we can figure out what path it followed to get there and where it's going next. A space vehicle could plot a course and know that at a certain time, at certain space coordinates, it would intercept the planet Saturn.

Early in our century scientists thought atoms were something like miniature solar systems with electrons orbiting the nucleus as predictably as planets orbit the sun. That made for excellent science fiction possibilities – our solar system as an atom in a superbeing's thumbnail – intelligent beings living on electrons, as we live on the earth. You were not taught in school that any such

13

possibilities existed, but you probably had a diagram of an atom in your physics book that looked something like the solar system, and you very likely carry that picture around in your mind even now to trundle out whenever the word 'atom' is spoken. In the 1920s physicists found that this is not an accurate picture of an atom (which shows the time lag between scientific discoveries and textbook publication). Though no mental picture really suffices, we do better to visualize electrons blurred in a cloud around the nucleus. With this revelation, science outdid science fiction.

As far as anyone has been able to discover, unlike a planet in a solar system, an electron (and this applies to all other particles as well) never has a definite *position* AND a definite *momentum* at the same time. We may measure very precisely the position of a particle, but we cannot at the same time measure very precisely its momentum. Or we may choose to measure its momentum very precisely, but we cannot at the same time precisely measure its position. It's as though the two measurements – position and momentum – ride at opposite ends of a see-saw. As the precision of one measurement rises, the other inevitably goes down, and vice versa. This is the Heisenberg Uncertainty Principle of quantum physics. No-one has been able to find a way around it. There probably is none. We cannot under any circumstances find out where a particle is AND the speed and direction of its movement. The answer to that question with regard to any individual particle at any given moment in time seems not simply unknown – not simply unknowable – non-existent.

There are a few physicists who still refuse to believe that such a bizarre situation, such a block to our further investigation, should be the end of the story. They hope that some future development in physics will increase our understanding and make it possible to ask and answer the two questions precisely at the same time: 'Where is the particle and how is it moving?' But most have concluded that this question has no answer, that quantum uncertainty does not result from our ineptitude as observers, that things on the quantum level really are uncertain.

The drama of this situation may not strike you immediately. It's obvious that no scientist likes to be frustrated in his or her investigations, but why should this uncertainty have so profoundly disturbed the scientific community when it was discovered early in our century, and in the years following that discovery? It has been disturbing in part because it seems to undermine our faith in the reliability of cause and effect, a concept which has traditionally supported the assumption of a rational universe.

In the case of the *Challenger* explosion, we had a definite series of events, a history (though perhaps not entirely known to us) that happened in one way and not in any other. In the case of the planets in the solar system, a definite path, a definite history brought Saturn to the position in which we observe it tonight. Even in the case of an amnesia patient who remembers nothing at all, whom no-one else can identify, and whose past cannot be traced, the patient is *assumed* to have a history which happened in one way and not in any other.

In the case of an individual elementary particle, that definite series of events, that definite history, is missing. The particle doesn't even have an unknown or an unknowable history. What it has is a blur of possible histories, a blur that does not focus itself on one historical track rather than another. Studying the quantum world from our level, we see that some histories for a particle are more likely than others, have a greater probability. Nevertheless (to state the case in its most extreme form) any history is possible and there is *no* answer to the question 'What history brought this one particle to the position or momentum we, at this moment, measure it to have?' In this sense, 'causality' is lost.

In case you are thinking that all this, though fascinating, is not very relevant to the world of everyday existence, let me remind you of the chair with which we began this chapter. All ordinary matter in the universe is made of atoms. That goes for this book, ourselves, planets, air, microbes, as well as chairs. Every atom consists of particles, and the uncertainty principle applies to all particles. You and I and chairs and tables and all other matter in the universe are at one level a quantum blur – on any level an amalgam of uncaused events!

But does a loss of causality on the level of particles and atoms really call into question the rationality of the universe? You may raise your eyes from this book and reassure yourself that day does follow night and night follows day, the seasons come round as expected, the moon and the planets keep to their appointed orbits, the galaxy retains its shape, the room in which you sit has the same dimensions it had an hour ago. Whatever nonsense is going on, it all sorts itself out into the familiar and, given the circumstances, surprisingly dependable world we perceive. In Chapter 6 we will examine reasons why this is so. But scientists still don't completely understand how and why the sorting out takes place, how and why the world of quantum uncertainty is transformed into the common-sense world of our daily experience. They cannot tell us how large a part human perception and consciousness play in the sorting out, how much 'interpretation' by the human mind has to go into the transformation, how much what we see is what we expect to see rather than what is really there.

We know we aren't directly conscious of everything that goes on around us. Our five senses are our only contact with the world, and there is much news they don't transmit. In the room with us there are many types of electromagnetic radiation that we aren't aware of. All of these are forms of light which human eyes can't see. Some of them we sense as heat. Others not at all. Some are in the form of radio waves, which we can't know are there unless we turn on a radio. What else is going on around us? Suppose the universe is really a place of nonsense – anarchical, meaningless, patternless, directionless in both space and time. Is there a possibility *that* is what reality is like? If so, why do we see so much pattern?

The theory of evolution tells us that certain capabilities give certain individuals within a species a survival advantage. These individuals are more successful than others at making the best of the situation in their environment; they live long enough to have more offspring. Their traits, including those which gave them the survival advantage, are passed down to more descendants than the traits of those individuals without the survival advantage. We'll discuss evolution in more detail in Chapter 6. Meanwhile, we are all probably familiar with examples of survival advantage. If lizards appear in green and brown, and green is a good camouflage among leaves – so that the predators of lizards can't find the green ones to eat them – after a few generations (all other things being equal) brown lizards are likely to be extinct and green lizards flourishing.

It isn't hard to imagine that in the evolution of living beings there was a survival advantage for those who could discover pattern in their surroundings and experiences. Brains would have evolved in such a way that as generations passed they were better and better able to find such pattern. We know that the human mind has become a superb device for compressing the wealth of information it receives from the five senses into useful, meaning-ful, abbreviated form. Thought and memory could not work as they do if we were not equipped to do this compressing. It doesn't seem far-fetched to think that our brains, having been wired this way by evolution, might continue with this process out of habit, even to the extent of finding pattern where there is no inherent pattern to be found.

But could the human or pre-human brain have created the very concept of pattern if there had been no pattern at all to be found in the universe? Is that perhaps how we have come to interpret a quantum blur as a chair? Are there in reality an infinite number of dimensions, only four of which our senses and our conscious-ness allow us to know about? Does time perhaps *not* flow chronologically in a way which allows us to remember the past but not the future? Can we prove anything about this at all? A good argument against an absence of all pattern is that evolution itself is a pattern. If *that* pattern exists only in our minds, could anything have done any evolving?

It's difficult to see how *all* pattern could be merely *our* invention. But could it be that human beings have come to attribute more importance to the pattern found in nature than nature herself does? Consider the symmetry we find in nature. We have only to look around us to see that there is far more to the picture than simple symmetry. Symmetry seems to be an ideal which much of the universe fails to live up to, at least on the levels most obvious to us.

When she was ten years old, my daughter did a school project about geometric shapes in the natural and built environment. Collecting photographs, she discovered it was easy to find examples in the built environment. Here were squares, pyramids, even dodecahedrons aplenty. Collecting the photographs of the natural environment was much more difficult. Circles were there in the pupils of our eyes and the ripples when we drop a stone in still water. But other shapes presented a problem. Columnal basalt formed *roughly* hexagonal shapes in a natural 'giant's pavement'. The hexagons in DNA spirals, in beehives, and in the compound eye of a horsefly also seemed carelessly drawn, without regard for exact geometry. The diamond shapes in a sunflower seed-head were lop-sided. One had to give tree-trunks the benefit of the doubt in most cases to call them cylinders. The earth bulges and is not a perfect sphere. Natural crystals are not perfect geometric shapes either. As for mirror symmetry, one side of a human face is not the true mirror image of the other.

It seemed at first to a ten-year-old that all the wealth of geometric shapes and figures that lie waiting in mathematics, which is arguably a thing of nature, not human-made, is largely unrealized in nature itself. Nature has not taken advantage of many of the possibilities. Humans have. The things we build and the art we create exhibit much more geometry and symmetry than we can find in nature. Are we bettering nature, imposing rationality on a less rational universe, when we design a building or draw a pleasing design?

Even my young daughter soon realized that the situation was really more subtle than that. There is geometry hidden in nature. The way we see, the way we judge distance and perspective is all bound up with triangles and cones. The frequency of vibration of the stretched membrane of a timpani head involves circles and wedges of circles. Radiation waves move out spherically from an underground explosion. The imaginary line drawn through time by a planet (not only its orbit, the pattern of many superimposed orbits, in time-lapse photography if you will) is beautifully geometric. The rules of geometry help dictate how a building can be made to stand and what cannot be built. Whether we choose to carry that geometry and symmetry into the more visual level of decor and design, we *must* adhere to nature's geometry-related rules in the structure. We shall see later that there is also symmetry hidden in the fundamental laws of nature.

But if symmetry and geometry go deeper than what we most readily see in the natural world, the digression from ideal geometry and symmetry also goes deeper. Matter in the universe, in the form of stars, planets and galaxies, is distributed unevenly. It is clumped in a way not yet understood by science, leaving enormous, mysterious voids. At the level of elementary particles, we discover a right- and left-handedness about the universe, slightly favouring the left. In the early universe there may have been an infinitesimal imbalance between the amount of matter and the amount of antimatter, an imbalance which has resulted in the universe of matter we see today.

If we were somehow able to take the natural world, straighten out the lines, correct the asymmetries and irregularities, make all the tree-trunks into true cylinders, the picture that would emerge would be unnatural, unbeautiful, impossible. If someone or something had taken the asymmetries found in physics and 'corrected' them, we and our universe could not exist. As important as the concept of pattern in nature is, there is also a powerful requirement for a pulling out of shape, a deviation from plumb, a tipping of the balance. There is a tension everywhere between ideal pattern and deviation from it. Can we call this tension itself a symmetry, a pattern, a balance? Such a subtle symmetry, such a tension, is familiar to artists and musicians. It is part of their craft to use it, to make it work for them. It is less familiar, perhaps, to scientists, except for those engaged in the study of chaos and complexity.

The rationality of the universe goes beyond its manifestation in obvious symmetry, pattern, and cause and effect. It would appear to include the ability to make judgements as to when the symmetry must be broken, when the geometry must be pulled out of shape, when cause and effect must not apply. Is that the rationality of the Mind of God?

Perhaps we have underestimated the amount of apparent asymmetry and 'irrationality' that can be accommodated without contradiction in a rational universe. Perhaps there is no contradiction between a rational God and a range of human experience that seems to stretch any conventional notion of rationality beyond the breaking point. Or are such suggestions merely a rearguard action engaged in so that we can preserve our assumptions that the universe is rational and that there is a Mind of God?

'IN NATURE'S INFINITE BOOK OF MYSTERIES . . .'
CAN WE READ VERY MUCH AT ALL?

'We don't know a millionth of one percent about anything,' said the American inventor Thomas Edison.[5] Of course it's been at

19

least sixty years since he uttered those words. We ought to know a little more by now. Even so, just a casual look around us would indicate that there is incredibly much to know.

We've already called into question the trustworthiness of our five senses, through which any information about the universe must come to us. Nevertheless we have a gut feeling (which isn't quite the same thing as common sense) that the universe is open to our study and our understanding, and this feeling certainly isn't new with our generation or our century. However, it is possible to conceive of a universe which would be rational yet somehow blocked off, veiled, difficult to find out about, as our everyday world must be to one who is born blind and deaf. It is even possible to conceive of a universe in which this blocking off would be for our benefit. T. S. Eliot wrote that 'Humankind cannot bear very much reality.'[6] Perhaps he was right.

Nevertheless, we yearn to know the truth about everything and behind everything, to see further and further with our telescopes, to probe closer and closer with our microscopes, to know all the answers. We are hard to discourage and not particularly humble in assessing our capabilities or our achievements.

In April of 1980 Hawking had the audacity to suggest we had come so far that before the end of the twentieth century we might find the theory that would explain everything that is happening, has happened or ever will happen in the universe. Eight years later he wrote that after we have that theory in hand we might just go on (not scientists alone, but all of humanity) to know the mind of God. Which calls to memory an ironic piece of history trivia. In the late 1890s Prussia closed its patent office on the grounds that all possible inventions had been invented. It wasn't long afterwards that Albert Einstein, in a Swiss patent office, began toying with ideas which would revolutionize science.

In the children's party game 'Pass the Parcel', a colourfully wrapped package goes round the circle of children while the music blares. When the music stops, the child holding the parcel unwraps the first layer of tissue paper. A piece of candy tumbles out, the reward for this child. The music begins, the parcel starts round once more, and the game goes on. With each pause in the music another layer of paper is pulled off and the parcel gets smaller. At the heart of the parcel there is a prize more exciting than any of the candy rewards that have come before.

Science plays a game like Pass the Parcel, unwrapping layer after layer of knowledge to reveal deeper knowledge, more complete understanding. For instance, unwrap atoms and you find electrons, protons, and neutrons. Unwrap protons and neutrons and you find quarks. Perhaps there are, after all, layers of structure more basic than electrons and quarks. As the game goes on, we hold our breath to see what will emerge when the last wrapping comes off. We might have to hold our breath for a very long time.

If our game is 'Infinite Pass the Parcel', it will never end. We will grow old and die sitting in that circle, listening to that tinny march! New generations will take our place in the circle. We will discover more and more refined theories, each one describing the universe more accurately than the last. Devise a way to take more sensitive measurements or make a new observation, and we discover things that are not accounted for by existing theory. Develop a more advanced theory. With each advance a layer of the parcel is unwrapped. The 'unknown' seems to become smaller. But if knowledge is infinite, the 'unknown' will never truly grow smaller. Every layer will reveal a deeper layer, and there will be another beyond that. Even if there is such a thing as complete knowledge, our way of doing science might mean that an infinite number of refinements would be needed for us to touch bed-rock. We may pass the parcel for all eternity. Einstein, for one, believed that 'this process of deepening the theory has no limits.'[7]

Whether or not nature's book of mysteries is infinite, science has already encountered some specific pages of the book which seem to be unreadable. We have already mentioned the quantum level of the universe and how the uncertainty principle limits us there. Physicist and author Paul C. W. Davies described scientific work on elementary particles as 'learning more and more about less and less.' Hawking calls quantum mechanics the 'theory of what we do not know and cannot predict.' Einstein didn't want to accept quantum uncertainty as inherent uncertainty. 'God does not play dice,' he declared. But Niels Bohr, the Danish physicist, who was convinced that the quantum world was intrinsically uncertain, answered, 'Albert, don't tell God what he can do!' In the 1930s Einstein devised an experiment which he hoped would show that events, even on the quantum level, have distinct causes. It wasn't until the 1960s that the technical capability was available to carry out Einstein's experiment. The results showed that Einstein had been wrong.

The quantum world does not provide the only unreadable passage in the book of the universe. For a time in the late sixties and seventies, it seemed as though singularities of infinite density and spacetime curvature might end all hope of our learning about how the universe began. If singularities exist, they are a serious road-block. Relativity theory predicts that we should find them at the centre of black holes, at the beginning of the universe, and possibly at the end of the universe. Physicists do not want to find singularities. It is no small matter to discover a door slammed in their faces.

First, a look at singularities which might be at the centre of black holes. Black-hole theory has it that a massive star, quite a bit more massive than our sun, after successfully supporting itself for millions or even billions of years against the inexorable collapsing pull of gravity, runs out of the fuel necessary to continue this support. To be more specific, the fuel is hydrogen, and the star has been producing energy by transforming this hydrogen into helium and then into some heavier elements. When the energy the star can produce is no longer enough to balance the pull of gravity, the star begins to collapse. If the star is massive enough, it will go on collapsing until it becomes a black hole.

What exactly is a black hole? The classical textbook definition is an area of the universe from which nothing can escape unless it is capable of travelling faster than the speed of light. Only the ability to exceed the speed of light could allow something to escape the gravitational pull of a black hole. Nothing that we know of can exceed the speed of light, and so it follows, by this definition, that nothing, not even light itself, can escape from a black hole.

If you aren't familiar with the concept of black holes you may be picturing an invisible solid sphere (the remains of the star) out in space, emitting no light and allowing no escape from its surface, but that isn't quite correct. A black hole is not an object but includes an area of space surrounding the collapsing star – roughly spherical but probably bulging like the earth does around its midriff. Relativity theory predicts that the star itself, within this area of no escape, goes on collapsing until all the matter in it is compressed to an area of zero volume and infinite density, which is known as a singularity.

Physical theories can't really handle infinite numbers. When Einstein's theory of general relativity predicts a singularity of infinite density and infinite spacetime curvature, that theory is also predicting its own breakdown. All the theories of classical physics break down at a singularity. We lose our ability to predict anything.

Some of you may be wondering why we don't label the entire interior of a black hole, rather than just the singularity, as *terra incognita*, one of the unreadable pages. If no light or anything else can come out of a black hole, then surely no information can come out. How do we know what goes on in there?

Black holes are indeed mysterious, but we do know mathematically and theoretically a great deal about them, including the dynamics of their interiors. Furthermore, it isn't entirely inconceivable that we may some day have the technology to travel to a black hole. Then if anyone is *really* curious, he or she can jump in, and if the black hole is large enough, so that gravitational tidal effects don't tear the explorer to spaghetti immediately, he or she can find out first-hand about what goes on inside a black hole, at least in its outermost areas. This expert witness won't be able to return to report on the experience to the rest of us, but at least one person's curiosity will be satisfied. The interior of a black hole is not unknowable.

However, it isn't the singularities that might lie at the heart of black holes that trouble physicists most. The really serious unreadable page is the singularity at the beginning of the universe. First we had better discuss why there should be any singularity there at all.

In the 1920s American astronomer Edwin Hubble made one of the most revolutionary discoveries of our century: The universe is expanding. The distant galaxies are all increasing in distance from us and from each other. If this is true, and no-one today seriously contests it, then unless something has changed dramatically in the past, the galaxies used to be much closer together. It follows that at some moment in the distant past everything that we might ever be able to observe in the universe would have been in exactly the same place. All that enormous amount of mass and energy would have been packed in a single point, infinitely dense.

We'll return to the events and controversies leading up to and following upon Hubble's discovery in Chapter 4. For the moment suffice it to say that although general relativity predicts the existence of singularities, it was not until 1970 that Roger Penrose of Oxford University and Stephen Hawking (both experts on black holes) used what they had learned from black holes, reversed the direction of time, and showed that the universe must have begun as a singularity. This was good news for their careers as physicists. In another way it was bad news.

If Hawking and Penrose were correct, the singularity at the beginning of the universe would mean that the beginning of the universe is beyond our science – an unreadable page. As is true at a singularity in a black hole, the laws and theories of classical physics, including Einstein's theory of relativity which predicts the singularity, all break down at the singularity at the beginning of the universe. We couldn't use these laws to predict what would emerge from the singularity. It could be any sort of universe. And the question of what happened before the singularity probably has no meaning at all. All we could say about the beginning is that time began, because we observe that it did.

It wasn't long before physicists, with Hawking in the lead, began to attack this ultimate Gordian knot. We shall see the results of that venture in Chapter 4.

There is another category of information about the universe which seems closed to our investigation. We have not yet found a way to predict 'constants of nature' such as the mass and charge of the electron and the speed of light in a vacuum. To say these are unknown would be incorrect. We can, in fact, measure the mass and charge of the electron and the speed of light. What we don't know about them is more subtle than that: If we couldn't measure these values directly, we wouldn't be able to find out what they are from any theory we have. These are 'arbitrary elements' in all our theories. An alien who had never seen our universe would have no way of finding out by using any present theory what these values would be in our universe. And that, for a physicist, is an unsatisfactory situation.

Will we ever know these answers? Some hopeful avenues are currently being explored. However, if our universe began as a wormhole tunnelling out of another universe, as one speculative theory we encounter in Chapter 4 suggests, we may never be able to predict all the constants of nature – though we will understand better why we must remain frustrated.

Relatively new branches of science called chaos and complexity, which we shall examine in detail in Chapter 6, lead us to believe we've been over-confident about human ability to predict even the orbits of the solar system very far into the future. With most, perhaps all, systems in nature, only infinite knowledge of present details (and perhaps not even that) would allow us to calculate precisely what will happen in the future of the system or what has happened in the past. We never have infinite knowledge of details. Where does that leave us in our gallant attempts to trace the history of the universe to its origin and to predict its future?

Chaos and complexity also point up a significant road-block between us and the fundamental laws of nature. When we try to understand the structure of the universe, we discover many instances where it is difficult, perhaps impossible, to determine whether what we see is the result of fundamental laws or the result of chance. If we are observing a chance outcome, one among many outcomes the fundamental laws would have allowed, then it would be misleading to suppose our observation is a clue to the fundamental, underlying laws of the universe. For example, if the way galaxies cluster is attributable to the laws of nature, we can study that clustering and learn something about those basic laws. On the other hand, if the way galaxies cluster is a matter of chance, with the underlying laws permitting a variety of results, we won't learn much about the basic laws by studying the way galaxies are clustered. It's a Catch-22. Not understanding what the fundamental laws actually determine, and where they are flexible, renders us incapable of finding out what those laws are.

IS OBJECTIVE REALITY A MIRAGE?

If we ask ourselves what we believe about the existence of objective reality, objective truth, the answer for most of us is probably that we think it exists, and we tend also to believe that science and the scientific method are the best way to get at it – to settle what is the truth and what is not.

However, science doesn't make any claim to have discovered the ultimate truth about anything. Scientists speak instead of discovering predictability – of seeking deeper understanding of nature. They don't speak of 'the verdict of science', but of 'the standard model', which means the model that nearly all experts agree on at the present time. They speak of 'approximate theories', which means theories that work satisfactorily in a certain area but do not claim to be the whole truth as it might apply to all areas. They speak of 'effective theories', which means something we can work with for the present while knowing that it isn't absolutely and unequivocally correct.

It is generally agreed that in science nothing can ever be 'proved'. The best anyone can say of a theory is that it has not been disproved. No matter how many times something is confirmed by testing, there is still an infinite number of times it may be tested in the future. That means the number of chances left for it to be disproved will always outnumber the number of times it has been tested and verified. Scientists are sceptical people when it comes to anything which claims to be ultimate, unassailable truth. It may be this scepticism that keeps some scientists away from a belief in God, not the notion that science disproves God. The idea of anyone actually finding ultimate, unassailable truth has in a sense become foreign to the minds of many scientists, and to some of the rest of us as well, even though we may believe such truth exists.

In other areas besides science, truth is even more elusive. Where questions of religion, morality, and human behaviour are involved, we are prone to say that it is a matter of opinion, a matter of belief. What happens to the notion of objective reality then? It is certainly very tolerant of Hawking, for example, to say that whether God operates in our lives is 'a matter of belief', but surely he doesn't mean that objective reality is different for the atheist from what it is for the person who believes in God. Does the Christian or Jew live in a universe that was created and is sustained by God and the atheist in a universe for which there is no God? If there is such a thing as objective truth, some of us are dead right and others dead wrong. Tolerance is necessary not because everybody is equally right, but because we have no way of proving once and for all which of us is right.

That is, IF there is such a thing as objective reality. It is not

26

inaccurate to say that on the quantum level of the universe the objective truth seems to be that we lose objective reality.

Recall the two ways we have mentioned of explaining the uncertainty we find on the quantum level of the universe. One way is to say that things there seem uncertain because we haven't yet found an adequate way of observing and measuring. However, the majority of physicists have become convinced that quantum uncertainty is something deeper than merely a matter of observation and measurement. When we measure precisely a particle's momentum, that particle does not at the moment of our measurement *have* any definite position to be measured.

That raises the question of whether anything that isn't located somewhere is a real 'thing'. Does it actually exist as an independent entity? If it does, wouldn't it *have* to have a definite location and a definite motion?

Even more troubling, there is a sense in which we as observers change reality on the quantum level. We said earlier that, as far as anyone has been able to discover, a particle or even an atom never has a definite position AND a definite momentum at the same time. If you look for an atom's position, that's what you get, an atom in a definite place – with a blur as to its motion. If you look for an atom's motion, that's what you get, an atom moving in a definite manner – with a blur as to its location. A very predictably unpredictable little fellow, this atom. But what happens when you aren't measuring anything at all about it? It seems that when an atom isn't being observed it lapses into a state that can be described as ghostlike, with no concrete reality to it at all. Only under observation does it resolve itself into either an atom with a location or an atom with a definite momentum, and which atom it will be depends entirely upon what the observer is trying to measure. To put it bluntly, the observer seems to create reality by observing it.

John Wheeler of Princeton and the University of Texas is the physicist who coined the name 'black hole', which was fortunate because the name 'collapsar' was the best anyone else had suggested. Besides inventing good names, Wheeler has a remarkable talent for finding analogies that make it possible for non-physicists to understand physics. Here is his version of Twenty Questions, Quantum Style.

Professor Wheeler is IT. We all assume that he has chosen a

secret word, but he decides to play a trick on us. He doesn't choose any word at all. The game begins. 'Animal, vegetable, or mineral?' we ask. Prof. Wheeler, having no secret word in mind, just a blur of every noun in his English vocabulary, is free to choose any of the three categories. 'Animal,' he answers. As we all shift our attention to the animal kingdom, the blur of possibilities becomes smaller. 'Mammal?' someone asks. 'No,' answers Prof. Wheeler, though he could just as honestly have answered 'Yes.' 'Reptile?' is the next question. 'Yes,' says Prof. Wheeler with a congratulatory nod, although he might just as truthfully have said 'No.' Now we all think of snakes and lizards and the like, a blur of reptilian life in our minds. A blur of reptilian life in Prof. Wheeler's mind too. There is no definite reptile lurking there in his mind's eye. As the game goes on Prof. Wheeler may have to be very clever in order to keep each answer consistent with all his previous answers, but if he does, can you see that in the end we will arrive at a definite word, although there was not one waiting to be found in Prof. Wheeler's mind? The avenue our questions have taken has helped create the hidden word.

In an analogous way, Wheeler tells us, it is our probing that determines what reality is on the quantum level. It isn't a reality that exists 'out there', independent of us, waiting to be found, the same regardless of whether or not anyone is looking. Our act of observation creates a real situation where otherwise there would be only ghostly uncertainty. We can't separate this reality from the person doing the observing or from his or her choice of how to do the measuring.

If we as observers manipulate and even create reality on the quantum level, what effect might we be having on the universe as a whole? It is Wheeler again who presents us with a mind-boggling suggestion. Perhaps it may be impossible for a universe to exist without observers. Does it follow that the universe did not exist before there were thinking beings in it? Does it follow that our observations create a history of the universe before our own appearance, a history that in a certain sense did not exist before we began to ask questions about the early universe? What meaning does our expertise and our technology have if all we are able to do with it is discover answers we are creating ourselves? And if we become extinct, will the universe vanish?

In a parallel line of thought, can God exist without believers? If the existence of God is a matter of belief, then if nobody believed in God we wouldn't have a lonely God, we'd have no God at all. Is it possible to conceive of a situation in which the answer to the question 'Is there a God?' is indefinite in the way particles are indefinite on the quantum level? Would belief then, not observation, create an affirmative answer? This would not be the same as saying the believer is deluded, any more than the physicist who locates an electron at a definite position is deluded. Unbelief would create a negative answer, and that also would not be a delusion. Can truth be contrary to truth?

Suffice it to say that most of us do not take kindly to the notion that it is possible to have contradictory truths. Contradictory opinions – all right. Contradictory evidence – that's OK too. Somebody's mistaken. Somebody's lying. Compromise – fine. But contradictory truth? No. Most of us feel instinctively that there is a definite answer to every question, even the question of whether God exists. We feel that our opinions and our beliefs do not make something real or unreal. We do not manipulate reality, whether that reality is the existence of a chair or the existence of God. In spite of hints to the contrary coming from the quantum level of the universe, when it comes to decisions about ultimate reality, I don't think my vote gets counted.

This is not a reaction confined to the ordinary, common-sense-oriented person-on-the-street. Most scientists feel there must be something 'real' or else what they study about the physical world would not fit together in such amazing and unexpected ways. We hear almost identical words from people regarding their belief in God. But isn't this 'fitting together', which we interpret to mean that there is some raw material out there against which we can stub our toes and bang our heads, just the sort of pattern that evolution has so superbly conditioned us to find – and to feel good about finding? Perhaps we even consciously or unconsciously single out problems for our scientific study which are likely to have that sort of satisfying resolution, while ignoring those which do not.

ARE WE REALLY FREE AGENTS?

A friend of mine, Jim Morgan, tells me that on a summer day in 1990, as he sat in a camp chair in his garden reading *A Brief History of Time*, a yellow leafhopper about a quarter-inch long landed at the top of page 9 and remained there for about six seconds. Jim stopped reading to ponder. Was it determined irrevocably at the instant of the creation of the universe 10 to 20 billion years ago that he and the leafhopper and Hawking's page 9 would meet, precisely thus, on this quiet summer afternoon? Not a second earlier, not a second later, not on page 8 or 10? Hawking, of course, has said he is of the opinion that everything that happens, has happened, or will happen in the universe *has* been determined either by a Theory of Everything or by God. Jim Morgan says he would like to think Hawking is right. If he is, any assumption that chance and choice play a role as events unfold in the universe is a false assumption.

What is a Theory of Everything? It isn't really correct to say *a* Theory of Everything. That would imply that there is more than one such theory. It must be *the* Theory of Everything – the simple set of rules that would underlie all the enormous complexity and trivial detail of the universe. A formula that could be written on a T-shirt? Maybe.

It isn't easy for a non-physicist to see how such a formula could exist. A glance out of almost any window or the thought of the working of our own bodies is enough to tell us there are far too many things going on in the universe to be explained so succinctly. But scientists have for centuries been finding that nature is often less complicated than it first appears. Richard Feynman, the American physicist and Nobel laureate, describes the way the process works. There was a time, he reminds us, when we had something we called motion, and something called heat, and something else again called sound.

> But it was soon discovered [Feynman writes] after Sir Isaac Newton explained the laws of motion, that some of these apparently different things were aspects of the same thing. For example, the phenomena of sound could be completely understood as the motion of atoms in the air. So sound was no longer considered something in addition to motion. It was also

discovered that heat phenomena are easily understandable from the laws of motion. In this way, great globs of physics theory were synthesized into a simplified theory.[8]

Hawking, in his inaugural lecture as Lucasian Professor of Mathematics at Cambridge, suggested that we may soon be able to synthesize *all* of physics theory into one simplified theory, but he was not suggesting that we will soon have a theory with which we human beings can predict everything that happens in the universe. We won't be able to use it to decide which horse to back in the Grand National. There are too many billions upon billions of details involved in tracing the history of every particle that makes up every horse, the turf on which it is running, not to mention the weather, from the instant the universe began to the day of the race. We have no computer capable of doing such calculations. There are other insuperable problems with predicting everything. Hawking thinks that's for the best. Otherwise, we'd place our bet and change the odds! Even our reaction to our prediction and the repercussions from our reaction would have to have been predicted in the theory.

Hawking was suggesting something less dramatic. He said that physics was well on the way to finding a theory which would give a unified explanation of the activities of the elementary particles and the working of the four forces by which they interact. These interactions underlie everything that happens in the physical universe. Hawking said in his inaugural lecture that the complete theory to explain the universe would also have to answer the question of what the 'initial conditions' of the universe were, conditions at the instant of beginning, before any time whatsoever had elapsed. We will see that a complete theory may have to do more than that, depending partly upon whose definition of Theory of Everything you are using.

However, the question we are asking here is not whether we can find such a theory, or what we humans could or could not predict with it, but rather does such a complete theory exist either within reach of us or beyond our comprehension. And if it exists, does it only *explain* everything, or does it actually *predict* or even *determine* everything? Is free will an illusion? Do chance and choice simply not exist? Perhaps even the Theory of Everything could not possibly have been different.

Although Hawking has said he believes everything is determined, he has also said that free will is 'a good approximate theory of human behaviour'.[9] We defined 'approximate theory' above as a theory which is useful in a limited context, but which may not be correct in all contexts. What Hawking means is that whether or not everything is determined, we do best to assume that we have free will and choices. And that's what most of us do. Even those with a strong belief in predestination still look both ways before crossing the street. Of course, one might argue that these people were predestined to look both ways.

Hawking is not alone in his belief that everything is determined, though there are probably fewer scientists who would agree with him today than there were in the eighteenth and early nineteenth centuries. The difference in Hawking's theories, as we shall learn in more detail later, is that they can be seen to undermine even the more fundamental assumption of contingency, that choice and/or chance were involved in the origin of the universe.

Other current science presents us with a different picture. Chaos and complexity studies reveal a delicate balance in the universe between predictability and unpredictability, allowing us to understand better why it is that we experience both in the common-sense world. We must save our discussion of chaos and complexity for Chapter 6 and a different context. For now, suffice it to say that they cast a strong vote against determinism, encouraging us to keep our assumption of contingency. However, there are also hints in chaos and complexity that the question 'Is everything determined?' can never be answered definitively by human beings.

The question of whether or not everything is determined appears repeatedly in science and religion and has profound implications for human morality. On the one hand, God is supposed to have foreknowledge. On the other, we are told we have free will and will be held accountable for our actions. How can both be true? On the one hand, the Theory of Everything may have determined the future from the instant of the beginning of the universe. On the other, we are told we do best to assume we live in a contingent universe, a not-entirely-predictable universe – a universe that can be studied only by looking at it, not by pure thought, no matter how advanced and well-informed that thought. The enormous paradox that lies at the heart of Western religion seems to lie at the heart of science as well.

IS THE UNIVERSE A UNI-VERSE?

The assumption that the universe has a unified description is less easy than the other four assumptions to support from everyday experience, which often seems to imply the opposite of unity. It might appear that only by limiting our scientific inquiry to what does fit into a unified picture could we possibly go on claiming such unity exists. Similarly, there are those who insist that only by shutting our eyes to contradictions and conflicting claims can we sustain a belief in one God.

Nevertheless, in science, our faith in this unprovable assumption of unity keeps us searching for deeper, simpler explanations in which the fragmented picture resolves itself into something of great simplicity, elegance, and beauty. If 'laws' break down, then what we have been calling 'laws' must be only approximations and we must look further beyond them for those laws which are truly fundamental and unchanging – an underlying symmetry. This way of proceeding has indeed proved fruitful. 'Beauty' is a strong guide in physics – a beauty which has to do in part with this falling into place of previously disparate elements. As our understanding deepens, contradictions often do seem to resolve.

Often . . . but not always. In mathematics, that area of thought where we most expect completeness and a relatedness without contradictions, we find contradictions. The mathematics worked out one way leads to one conclusion, and worked out another leads to a contradictory conclusion. We have learned to trust mathematics as a guide to what the real world is like – all of us in simple ways, theoretical physicists in ways they think will lead them to fundamental understanding of the universe. Could it be that our mathematics sometimes builds houses of cards? Or should we give the strongest interpretation to the way mathematics always seems to match nature and conclude that if there are contradictions in mathematics, there are contradictions in nature? What happens to our unity then?

Our assumption that there are laws which hold at all times and in all places leads us to believe that by studying a small part of the universe we can make great strides toward understanding the whole universe and its entire history – even predict its future. When the breakdown of physical laws at a singularity called into question the assumption that unchanging laws held even at the origin of the universe, this provided a strong motive to look for theories which undermine singularities. But if we favour theories which uphold our assumption of unity, do we risk a circular argument, letting our assumption pick our theory while our theory upholds our assumption?

What shall we conclude? Can we learn anything meaningful about the universe by means of science? Are not the assumptions which underlie the scientific method called into question by twentieth-century scientific theories and discoveries? Should we trust even those theories and discoveries? Haven't they also emerged from a structure which may be no more than a dubious inheritance from seventeenth-century religious dogma?

It may be an act of faith alone, a flying in the face of some contrary evidence, but few of us would succumb to complete pessimism or abandon the scientific quest. Few of us would say that the human race and individuals among the human race can't know anything meaningful about the universe. Some of us do go on doing science, and others search for God, and still others do both, or keep their options open, or merely cope on a day-to-day basis – continuing to assume that the universe is rational, contingent, open to our scrutiny, has underlying unity, and that there is such a thing as objective truth. Beyond that shared mind-set, we are a diverse and rather motley crew, like knights on a quest with many different motives and hidden agendas and varying degrees of commitment. In the chapters to follow we shall see where this adventure has led us so far, and where it might still take us.

3

ALMOST OBJECTIVE

> Experience is not all, and the savant is not passive; he does not wait for the truth to come and find him, or for a chance meeting to bring him face to face with it. He must go to meet it, and it is for his thinking to reveal to him the way leading thither. For that there is need of an instrument; well, just there begins the difference . . .
>
> HENRI POINCARÉ[1]

IN THE FILM *MONTY PYTHON AND THE HOLY GRAIL* A SCURVY BAND of knights went on a quest riding imaginary horses. The film budget was too small to afford the real chargers that would have carried these heroes in proper style. But no matter, imaginary horses were exactly right for this adventure. In a universe seemingly bereft of any clear concept of 'horse', the knights tried to enter an impregnable fortress by building a giant wooden rabbit in which to conceal themselves.

The end of Chapter 2 reminds me of those knights. It also brings to mind a recent Harry Kupfer production of Richard Wagner's *Ring of the Nibelungs* at the Bayreuth Festival in Germany. Wagner's gods and heroes inhabited a world lit by laser light, and their spears, swords, helmets, shields, even the luggage they brought with them to move into Valhalla, were all made of transparent plexiglass. Shafts of light ricocheted off the see-through weapons and suitcases full of nothing visible to human eyes, and swept into the corners of the dark Festspielhaus, occasionally spotlighting and blinding for a moment one or another of us in the audience.

Given a choice between the two images to symbolize humanity's quest for truth, I would prefer Harry Kupfer's clarity and piercing light over Monty Python's mud, blood, and humour. Of course, as Richard Wagner tells the tale, the magic weapons of Valhalla are in the end as ineffective and pitiful as Monty Python's giant wooden rabbit – and the horseless knights just might after all stumble on the Grail. Nevertheless the clarity and light looked more promising.

'There is need of an instrument . . . just there begins the difference . . .' says Henri Poincaré in the passage quoted at the head of this chapter. We who live at the end of the twentieth century have a lot of faith in the 'instrument' we call the scientific method. Here, we like to think, are clarity and light. Light to pierce the darkness of ignorance; clarity of vision for discerning truth. However, if we carry the Bayreuth metaphor a little further we discover that light also blinds us and causes deep shadows, and what is transparent sometimes becomes a mirror reflecting ourselves.

Be that as it may, how can we *not* have faith in science when it has produced the technology that underpins our modern civilization? Don't we believe the lamp will go on when we flick the switch . . . the microwave will cook our casserole . . . our fax will go through via satellite to Perth? Haven't we travelled to the moon and back? Aren't our space telescopes exploring the furthest limits of the universe? Aren't our instruments probing the fundamental structure of matter and revealing the deepest secrets of heredity and organic life? Certainly there are annoying glitches now and then. Certainly we have serious reservations about some aspects of the lifestyle our technology makes possible. But there is no doubt about it, science is a mind-boggling success story.

Nevertheless, should we accept everything that science tells us?

Science has never asked us to do that. Faith in the dogma-of-the-leading-edge-of-science wasn't invented by scientists – not the faith which blindly embraces the latest findings as the best findings, the right findings. Not the faith that speaks knowingly about 'what modern science tells us' or 'the verdict of science' and fails even to note whether it's talking about speculative theory and preliminary findings or well-established scientific knowledge.

It's the inalienable right of science to be wrong, and it insists on this right by putting its most cherished assumptions mercilessly to the test. Some like to think of science as chipping away at the universe to reveal Truth, as Michelangelo chipped away at marble to release a human form. In fact a large part of what science chips away at is itself. No, we can't believe everything we hear from science. We can only believe, as most scientists do, that through this chipping process eventually the truth will out.

Even so . . . can we expect science to provide all the answers? Will it lead us some day to ultimate and complete truth about the universe and beyond?

Now that is a different matter entirely. There are problems we can tolerate when we're searching for partial truth and predictability but which become daunting obstacles when our goal is ultimate and complete truth. How are we to find a way of looking at the universe free from any bias whatsoever? How are we to recognize ultimate truth when and if we find it? How are we to prove we've found it? The naive view is that these are precisely the problems science solves: it enables us to study reality without a bias, with pure objectivity, and prove conclusively what is true and what isn't. That is a *very* naive view of science.

Think of this chapter as a briefing session. In Chapter 4 we'll put ourselves in the hands of science and allow it to take us to explore extremes of space, time, and scientific imagination, assaulting strongholds once reserved for religion and philosophy. In preparation, though we can't all learn to think like Einstein, Hawking, and their colleagues, we must try to equip ourselves a little better to think *with* them – and that means having a more sophisticated conception of the way science works.

WHERE IS FANCY BRED?

The foundation stones of science are the assumptions we discussed in Chapter 2: the universe is rational, contingent (subject to chance and choice), accessible to human minds, has unity, and there is such a thing as objective truth. Beyond those foundation stones, what we learned in school about the scientific method can be reduced to two basic principles:

1. All our theory, ideas, preconceptions, instincts, and prejudices about how things logically ought to be, how they in all fairness ought to be, or how we would prefer them to be, must be tested against external reality – what they *really* are. How do we determine what they really are? Through direct experience of the universe itself. This assumes of course that we all know what we mean by 'direct experience of the universe' – but leave it at that for now.

2. The testing, the experience, has to be public, repeatable – in the public domain. If the results are derived only once, if the experience is that of only one person and isn't available to others who attempt the same test or observation under approximately the same conditions, science must reject the findings as invalid – not necessarily false, but useless. One-time, private experience is not acceptable.

What *is* acceptable from private sources is suggestions about what *might* be true, and here some creativity comes into play. Einstein wrote: 'When I examine myself and my methods of thought, I come close to the conclusion that the gift of fantasy has meant more to me than my talent for absorbing positive knowledge.'[2] Hawking has said: 'The ability to make these intuitive leaps is really what characterizes a good theoretical physicist.'[3] Poincaré calls it the ability to 'make ascensions otherwise than in a captive balloon'.[4] There is art to this science.

Creativity isn't limited to the theoretical side of science. If independent reality did come out to meet us or lie waiting to be found in an unambiguous fashion, we might justifiably conclude that creativity among observers and experimenters would be unnecessary and undesirable. We wouldn't want human subjectivity intruding between us and direct experience of the universe. But nature isn't that obliging. Scientific findings have to be coaxed out, dragged out, tricked into appearing, chipped at, hammered out, both mined and honed. Poincaré was right to say, in the passage at the head of this chapter, that if we want to find truth it is for our thinking to reveal to us 'the way leading thither'. When we choose or devise a 'way leading thither' we inevitably adopt or impose a point of view.

That was a dilemma Darwin faced when he collected data in

the Galapagos Islands. As science historian John Hedley Brooke tells it,

> Because [Darwin] stated in his private journal that the species of the Galapagos had marked the origin of all his views, popular accounts . . . have conjured up the image of a patient fact collector suddenly bowled over by what he had found . . . The trouble with such accounts is that they can trivialize the logic of discovery. They assume that the 'facts' were somehow there, waiting at the Galapagos for Darwin to process. Darwin himself knew better than that. One of the things that had worried him earlier in the voyage was whether he was noting the right facts. What his experience at the Galapagos rather embarrassingly showed was not that some new facts pointed unequivocally toward a new theory, but that the constitution of a relevant fact depended on prior expectation.[5]

It's never a simple matter talking about experiencing independent reality. If we ever felt we knew exactly what that meant, our confidence has been severely shaken by the quantum level of the universe. On that level we find that the concept of the scientist as a spectator, detached from what he or she studies, isn't viable. As we saw in Chapter 2, what's real there seems to depend upon whether we observe it and how we observe it. There is apparently no fundamental quantum reality in the sense we usually mean 'reality' – something whose existence is the same whether we're observing it or not, waiting to be discovered and studied.

In part because of our encounter with the quantum situation, we are more than ready to suspect that on any level of the universe how we look changes what we find. It becomes of paramount importance to ask which point of view it is we're operating from, how we've arrived at this point of view rather than some other, and how much it limits us. As the German physicist Werner Heisenberg said, 'in science the object of research is no longer nature itself, but man's investigation of nature.'[6]

'How we look' can mean anything from which apparatus we decide to use in the lab on Tuesday to how we allocate the national science budget over the next decade.

On a day-to-day basis, there are choices among techniques or equipment for carrying out an experiment, choices about which data are significant for a hypothesis, which of the more difficult-to-get-at data deserve special efforts and which doesn't, which piece of inconsistent data it is safe to ignore and which it isn't. On a broader scale, scientists opt for one theoretical or mega-theoretical framework over others, and allow that framework to direct the course of research. In a large or small way, each of these decisions is a decision about how we look.

Points of view also come from sources we think of as having less legitimate right than scientific technique or theory to influence what we find with our science, things more insidious and harder to control: individual preference, cultural conditioning, religious or anti-religious belief, political and economic interests, our value system, the spirit-of-the-times, the current fads in science. There's an almost inevitable tendency to find a theory more plausible if it's congenial to current thinking, both inside and outside the science community.

Other points of view are not insidious but surprising. Our vision is circumscribed by the fact that we live in the universe when it is ten to twenty billion years old and not at another time, on the earth and not elsewhere, and with the brains we have, capable and perhaps over-ready to compress information to patterns. With all our genius and scientific imagination, can we even begin to perceive how different we might find things if we could look from another time or place, with other senses and minds, or if we could see all of it? We are, as physicist Murray Gell-Mann put it, 'such a small speck of creation believing it is capable of comprehending the whole'.[7]

The scientific method itself is a point of view. Its assumption of order and search for an orderly universe have been with us since the seventeenth century, spotlighting areas which hold most promise for systemization while avoiding those which don't. British astrophysicist John D. Barrow writes in his book *Theories of Everything*, 'We [had come] to think of linear, predictable, and simple phenomena as being prevalent in Nature because we were biased towards picking them out for study.'[8] All five assumptions we discussed in Chapter 2 influence what theories we find most acceptable. So does our view of mathematics. Furthermore, the scientific method limits itself to evidence which can be corroborated.

Experiencing independent reality unentangled by any point of view or bias – with an unlimited range of experience . . . that would be discovering the chair-as-it-is-in-itself. With all we've learned, we still fall far short of that.

Can our point of view affect what we find? You don't have to believe things are as uncertain on all levels as they are on the quantum level to see that it can. Nor do you have to believe that a point of view changes objective reality. The choice of an experiment that is more likely to coax out one set of evidence than another; the choice based on a theory as to which evidence will be more significant and ought to be coaxed out; the choice of which theory we ought to take seriously . . . such choices don't change objective reality, but they do help determine what we *perceive* as reality and what will emerge as scientific knowledge. Some critics believe the result is 'knowledge' that has no basis whatsoever in objective reality. Should we treat such scepticism seriously? Enough so, perhaps, to take a really good look at 'how we look'.

Theory plays an enormous role in modern science. Particularly in physics, it provides the point of view and plots the course. This is by no means to discount the importance of observation and experiment, or to suggest that nothing theory hasn't anticipated comes from them. Plotting the course doesn't mean ignoring the terrain. Theorists frequently adjust theories to make sense of new findings, even of complete surprises, and when theories compete they opt for the one which is more consistent with experimental and observational evidence. Science doesn't run rough-shod over data. However, it's theory which assimilates the data, and after appropriate revision continues to plot the course.

That being the case, where do we get theory in the first place?

The quotations above from Einstein, Hawking, and Poincaré about fantasy and leaps of intuition might lead one to think a theory can be a complete flight of fancy. Is it really fantasy and intuition that plot the course of science?

It's true that a theory can go as far beyond previous scientific knowledge as a theorist's imagination can take him or her. However, most theory is built by reasoning on the basis of previous scientific knowledge, evidence, and other successful theory. All scientific theory, however it arises, is supposed to be logically consistent with such knowledge and evidence. If a theory

41

is not, it should explain the discrepancy or suggest what as yet undiscovered evidence we should look for which will show that our previous conclusions were off the mark. For example, after the planet Uranus was discovered at the end of the eighteenth century, astronomers found that its observed positions were difficult to reconcile with positions predicted by calculations using Newton's theories. However, the discrepancy could be accounted for if one figured in the existence of an as yet undiscovered body whose gravity was affecting Uranus' orbit. Newton's theory wasn't wrong. In fact it allowed astronomers to predict the position of the unknown body. In 1846 the unknown body – the planet Neptune – was discovered less than one degree away from the predicted position.

A strong theory gathers into the fold a broad range of evidence, making sense of what was previously unexplained, confusing, or contradictory. Supersymmetric string theory is a theory that sees the fundamental structure of the universe not as point-like particles (such as the electrons and photons we are accustomed to thinking about), but as tiny vibrating strings or loops of string. One argument in favour of superstring theory is that while other theories show inconsistencies when they attempt to incorporate gravity into quantum mechanics, superstring theory wouldn't be consistent if gravity *didn't* exist.

Other points in favour of a theory are successful interlinking with the network of existing theories; fruitfulness in giving rise to other theories and technology; and ability to avoid arbitrary elements. Arbitrary elements are things that can't be predicted by the theory itself, that must be taken as a given in order for the theory to work.

A criterion to which scientists attach great importance is how economical (the technical term is 'parsimonious') a theory is – does it refine ideas to a simpler, more self-evident form? This is not a criterion confined to science; it reflects an instinctive way of looking for explanations. We don't seek out a complicated explanation when there is a simple, self-evident one available. If we see a medium-sized, shiny black winged creature with a beak in the tree, we say it's a crow, we don't run for our book of exotic birds or speculate that it might be a bird previously thought to be extinct. If we hear a loud bang, we decide it was a car backfiring or a firecracker, only dimly entertaining the notion that our usually placid neighbour may have shot his wife. However, in science and in everyday life, it is only an assumption that the simplest, most economical explanation is most likely to be the right one – a problem which takes on major proportions if we want to prove to ourselves that we've found THE explanation for everything.

We expect a theory to make predictions for future testing. It isn't enough simply to tell us what to look for to show that the theory is correct. The theorist should also tell us what to look for that would show the theory is incorrect. Because of the nature of the theories we're going to be discussing later, we'd best concentrate for a moment on the requirement that there be ways to demonstrate that a theory is incorrect. At the outer reaches of scientific imagination, theories become almost 'un-falsifiable', and our next chapter is going to take us to those outer reaches.

There's a difference between failing to prove that something is true, and proving that it is false. Philosopher of science Karl Popper pointed out that no hypothesis can ever be proved by experiment. No matter how many experiments confirm the hypothesis, there is still an infinite number of chances remaining in the future for an experiment to produce different results. To say that something isn't proved is not nearly so strong a statement as to say it has been proved incorrect. 'Not proved' does not equal 'false'.

At the frontiers of physics, we find a number of theories which have very little hope of being tested by experiment or observation in the foreseeable future – if ever. The microscopic level at which wormholes may appear, the unstable primordial nothingness which might have decayed into something, the centre of black holes or the origin of the universe where we could search for singularities of infinite density, the era when time may have been a space dimension, the instant early in the first split second of the universe when gravity might have been a repulsive force – all are far beyond our observational and experimental reach. By one definition of the word, these theories are 'metaphysical'. Yet we will be making hypothetical journeys to all these places and times and giving all these theories a great deal of serious attention in Chapter 4. Why? Only because they haven't been proved wrong? Is *everything* believable that can't be falsified?

43

Yes. Technically, any idea that, if it were true, wouldn't overturn orthodox, successful scientific knowledge must be termed scientifically believable. But we've already seen that there's more to theory-making than devising a theory so safely far-out that it can't be falsified. In fact a theory with no possibility of falsification isn't considered a very strong theory. The proposals just mentioned are plausible to the extent to which theorists have shown they are mathematically and logically consistent within themelves and with known scientific laws and observational data, and they *could* be falsified by showing that they lack this consistency. Such distinctions are important when we're discussing highly speculative scientific theories, and they become particularly significant when we begin to ask questions such as whether belief in God is falsifiable. If so, by the standards of what laws and evidence? If not, is God not a very strong theory? And does that matter in the case of God?

We've said that theory is a legitimate channel for a point of view in science. It's a point of view we think we know how to handle, first because we're fully conscious it *is* a point of view – nothing insidious about it – and second because we have established ways of putting most theories to the test. But if a theory were to distort our perception of reality in a hidden way, in a way which skewed the fairness of the test itself, then admittedly we would have a problem. Do some theories do this? That's a little like the question I was asked on a survey: 'Do you unconsciously discriminate on the basis of race, class, or sex?' Once the question is raised, we can't possibly answer it with an unequivocal No. We have to consider the possibility: Could how we look (determined by a theoretical point of view) dictate what we find in such a way as seriously to skew scientific findings – without our knowing it?

THE SPECTACLES-BEHIND-THE-EYES

Russell Hanson, in a book written in the late 1960s, *Perception and Discovery,*[9] made the startling suggestion that scientific theory not only makes predictions that can be tested but sometimes also dictates what we discover when we do the testing

– helping to assure its own verification. Hanson calls theory the 'spectacles-behind-the-eyes' of science.

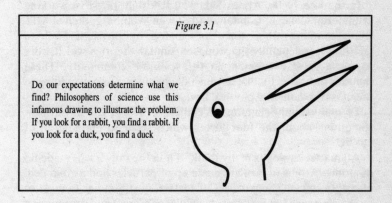

Figure 3.1

Do our expectations determine what we find? Philosophers of science use this infamous drawing to illustrate the problem. If you look for a rabbit, you find a rabbit. If you look for a duck, you find a duck

Spectacles are supposed to help us see what's there, and most of the time that's what they do. But spectacles can play tricks on us. I remember a set of 'spy' spectacles I had as a child. I ordered them by sending in a form on a cereal box. They came – two pairs of cardboard-and-cellophane spectacles and a card covered with dots like a Seurat painting. The dots didn't make a picture, but if you wore one pair of spectacles, you could see a faint message. It said 'Beware.' If you wore the other pair, the same card read 'Proceed.' If I had owned only one of the pairs of spectacles, I would never have discovered the second message. Had I owned neither, I would have discovered no message at all.

An example which some believe supports Hanson's suspicions about spectacles-behind-the-eyes cropped up, not long after he wrote his book, in the experimental evidence verifying the electroweak theory. This theory is one of the most significant advances in twentieth-century physics. It takes us a giant step nearer to the simplicity we think underlies the universe. In the next few pages we're going to spend some time with the electroweak theory, not only because it illustrates the point about spectacles and the reasons most scientists do not feel they lead us into error, but also because it's a good introduction to what we mean by a Theory of Everything and some other concepts we'll need in the chapters to come. The electroweak theory is not a Theory of Everything, but it's considered to be a move in that direction.

In the late 1960s, Abdus Salam, a Pakistani physicist working at Imperial College, London, and Steven Weinberg, then at MIT (both thinking along lines anticipated by physicist Sheldon Glashow), independently proposed similar theories which were to cause great excitement in the scientific community. These theories promised to take us much deeper in our understanding of the most fundamental physical laws.

To understand Salam and Weinberg's theory, you must know something about the four forces which seem to underlie all of nature:

All matter as we normally think of it in the universe is made up of atoms. Atoms in turn are made up of particles and a great deal of empty space. The particles of matter most familiar to most of us are electrons (which orbit the nuclei of atoms) and protons and neutrons (clustered in the nuclei). Protons and neutrons are made up of more fundamental particles of matter called quarks. All of these matter particles belong to a class of particles called 'fermions', named after the great Italian physicist Enrico Fermi. Fermions have a system of messages that pass among them, causing them to act and change in various ways. Think of a human society which has a message system consisting of four different services: telephone, fax, mail, and carrier pigeon. Not all the humans would send and receive messages and influence one another by means of all four message services. If you think of the message system which carries the messages among the fermions as four such services, which we call forces, you won't be far wrong. Other particles serve as messengers, and sometimes these also pass messages among themselves. 'Messenger' particles are more properly called 'bosons'. It seems that every particle in the universe is either a fermion or a boson.

One of the four fundamental forces of nature is gravity. A way of thinking about the gravitational force holding you to the earth is as 'messages' carried by bosons (in this case they would be 'gravitons') between the particles of the atoms in your body and the particles of the atoms in the earth, influencing these particles to draw closer to one another. A second force, the electromagnetic force, is messages carried by bosons (in this case 'photons') among the protons in the nucleus of an atom, between the protons and the electrons nearby, and among electrons. It causes electrons to orbit the nucleus. On the level of our everyday experience, photons show up as light, heat, microwaves, and radio waves. A third message service, the strong force, causes the nucleus of the atom to hold together. Its messenger particles are 'gluons'. A fourth, the weak force, causes radioactivity.

The gravitational force, the electromagnetic force, the strong nuclear force, and the weak nuclear force . . . the activities of these four forces are responsible for all messages among all fermions in the universe and for all interactions among them. Without the four forces, every fermion, every particle of ordinary matter, would exist, if it existed at all, in isolation, with no means of contacting or influencing any other, oblivious of every other. To put it bluntly, it would seem that whatever *doesn't* happen by means of one of the four forces . . . doesn't happen. That, when you think about it, is a very strong statement. If it's true, then a complete understanding of the forces would give us an understanding of the principles underlying everything that happens in the universe.

Much of the work of physicists in this century has been aimed at learning more about how the four forces of nature operate and how they're related. In our human message system, we might discover that telephone and fax are not two separate services, but the same thing, showing up in two different ways. That discovery would 'unify' the two message services. In a roughly similar way, physicists have sought to unify the forces of nature, hoping ultimately to find a theory which explains all four forces of nature as one 'superforce' showing up in different ways, a superforce which also unites both fermions and bosons in a single family. Such a theory would be a significant step on the way to a theory that would explain the universe – the so-called Theory of Everything.

Another ingredient of a Theory of Everything would be the 'boundary conditions' of the universe. If you set up a model train layout, position several trains on the tracks, and set the switches and throttles as you want them, before turning on the power, you are setting up boundary conditions. As far as this session with the train set is concerned, reality begins with things in precisely this position and not in another. Where the trains will be five minutes

47

later and whether they will crash depends a great deal upon these boundary conditions. Since these are boundary conditions at the beginning of the game, we call them initial conditions.

Suppose a friend comes into your train room ten minutes later. You kill the power. Now you have another set of boundary conditions – the exact position of everything in the layout at the moment you turned it off. You might ask your friend to figure out precisely where all the trains started out ten minutes earlier.

Scientific experiments have boundary conditions something like that – the lie of the land at a particular point in time, for instance the beginning of an experiment. Scientific observations and theories of the universe do too, except that we may have less choice of how to set them up. If I ask how many ways the universe could have begun and still end up the way we observe it today, assuming that the laws of physics as far as we know them are correct and haven't changed, I am using 'the way we observe it today' as a boundary condition. I am also, in a more subtle sense, using the laws of physics and the assumption that they haven't changed as boundary conditions. The answer I'm after is the answer to the question: What are the boundary conditions at the *beginning* of the universe, or the initial conditions of the universe – the exact layout at the word 'go', including the minimal laws that had to be in place at that moment in order to produce at a certain time in the future the universe as we know it today?

In addition to providing a unified description of the particles and forces, and boundary conditions for the origin of the universe, the Theory of Everything would have to be able to account for values which we said earlier are arbitrary elements in all present theories – including 'constants of nature' such as the mass and charge of the electron and the velocity of light in a vacuum. We know what these are from observation, but the Theory of Everything should explain and predict them.

If nature really is perfectly unified, then the initial conditions, the most fundamental particles, the forces which govern them, and the constants of nature may all be interrelated in a unique and completely compatible way, which we might be able to recognize as inevitable, absolute, and self-explanatory. When we speak of the Holy Grail of Science, the Theory of Absolutely Everything, it is that compatibility we mean – not just a complete description of the universe, but the answer to the question, Why

48

does the universe fit this description? With this goal in mind, we can see why an insight such as Einstein's, that gravity is not merely something affecting objects but must also be thought of as the warp of spacetime *caused* by the presence of objects, is more than just an interesting theory. It is an insight into a deep interrelatedness of nature.

Abdus Salam and Steven Weinberg's contribution to this quest for a final theory in the late sixties was the proposal that the electromagnetic force and the weak nuclear force are not two separate forces of nature but the same force showing up in different ways.

Salam and Weinberg knew that in the electromagnetic force, the photon (the messenger particle of that force) has no electrical charge of its own and doesn't change the electrical charge of the particles which send and receive its message. They thought it might be possible that some of the messages of the weak force also carry no charge. If so, then perhaps the weak force messenger and the photon are really identical twins showing up in disguise. The idea wasn't without problems. If this was a disguise, it was a very good disguise. The weak force messenger can travel only such short distances as exist in an atom, while the photon can travel for any distance in the universe at the speed of light. Nevertheless, Salam and Weinberg felt that the two particles (the massive weak force messenger and the massless photon) might appear in an identical way in the underlying equations.

Salam and Weinberg's proposal may seem, on the basis of common sense, not a ridiculous idea but perhaps also not a very promising one. However, science is full of instances where what appear to be totally dissimilar situations actually reflect the same underlying laws. Who at first glance would think that the same force which causes a ball, thrown upward, to return to the earth also causes the planets to orbit the sun in elliptical orbits and has prevented the universe from expanding so steadily that life as we know it could never have emerged? One of the problems scientists have to deal with is that simple underlying laws tend to manifest themselves in confusing and contradictory ways in the world we are able to study.

It often happens that something symmetrical in the underlying physics comes disguised as something not symmetrical. The solution which Salam and Weinberg suggested had to do with this concept, known as symmetry-breaking. We're using the word symmetry here in a way that may be new to some, but I believe it will be clear as we proceed. We begin with underlying laws which are symmetrical, meaning that they make a number of outcomes (manifestations of the laws in the evidence we can observe) equally likely. But none of these equally likely results is symmetrical. A simple example is a rod set on end. By law it's allowed to fall in any direction. The force – gravity – which makes it fall is symmetrical in that it doesn't prefer that the pole fall in one direction rather than another (see Figure 3.2). All the outcomes (the directions it could fall) are equally likely. But it

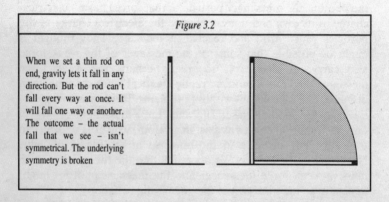

Figure 3.2

When we set a thin rod on end, gravity lets it fall in any direction. But the rod can't fall every way at once. It will fall one way or another. The outcome – the actual fall that we see – isn't symmetrical. The underlying symmetry is broken

can't fall every way at once. It *will* fall one way or another. The outcome – the actual fall that we see – is not symmetrical. We say the symmetry is broken. Another example: A magnet doesn't become a magnet as long as it remains above a certain temperature (see Figure 3.3); above that temperature, the forces acting on the atoms in the metal don't have any preferred direction. One direction is as good as another, the situation is symmetrical, and the bar of metal has no overall magnetism. Below the critical temperature, the atoms orient themselves in one direction. It could be any direction, but it can't be every direction at once. As the atoms orient themselves, the symmetry is broken and we have a north and south pole to the bar of metal. It is a magnet.

Figure 3.3

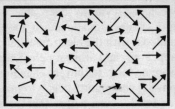

Above a certain temperature, the forces acting on the atoms in a bar of metal don't have any preferred direction. The situation is symmetrical, and the bar of metal has no overall magnetism

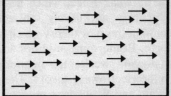

Below the critical temperature, the atoms orient themselves in one direction. The symmetry is broken and the bar of metal is a magnet

Surely one of the most intriguing examples of symmetry-breaking has to do with the direction of time. With very few exceptions, the laws of physics are symmetrical with respect to time, meaning that they work equally well forwards in time and backwards in time. They don't prefer a direction. You could make a film of most physical interactions and reverse the direction of the film and no-one who saw it could say which way it ought to run. But we all know that the outcome of this underlying physics in our universe is not time-symmetrical. For some reason we have a well-defined future and past. It would be difficult to mistake one direction for another. How this symmetry-breaking occurs is still one of the great mysteries.

Salam and Weinberg's use of symmetry-breaking in their theory was to propose that at very high energies, such as were present early in the first split second of the universe, the photon and the weak force messenger were identical twins. The situation was symmetrical. At lower energies, such as are present in the universe today, the symmetry is broken. The particle is either a massless photon or a massive weak force messenger. The fact that they are really identical twins is a secret hidden in the underlying physics. Before Salam and Weinberg arrived at this insight, physics had been suffering from a point-of-view problem: We live in an era in the history of the universe when such deep symmetries of nature are very long since broken.

The weak force, which no-one had previously been able to explain in an entirely satisfactory manner, made much more sense in this new theory which unified it with the electromagnetic force. Bothersome infinities disappeared, and arbitrary elements in previous weak force theories were no longer arbitrary elements in the electroweak theory. The physics community began to take Salam and Weinberg's proposal seriously, even before there was any experimental verification whatsoever.

Experimental evidence was not long in coming. Among other predictions, the electroweak theory predicts something called a neutral current in the operation of the weak force – the way the message is carried without the exchange of any electric charge (as the photon does in the electromagnetic force). In the early 1970s experimenters at the European Centre for Nuclear Research (CERN) in Switzerland and Fermilab near Chicago discovered just such neutral currents. As a result, physicists were, in general, convinced that the electroweak theory was correct, and a Swedish newspaper even predicted that Salam and Weinberg would win the 1975 Nobel Prize. So far, this appears to be a textbook example of the scientific method at work: Theory predicts – experiment tests and (in this case) confirms.

However, it is unsettling to note that experimenters *could* have found the neutral current in the early 1960s, before Salam and Weinberg proposed their theory. The evidence was there. We might say experimenters did find it, without recognizing it for what it was. The neutral current showed up back then in experiments with the weak force – 'showed up' in the sense that physicists noticed things going on which other experimenters later would explain were due to neutral-current effects. However, there were many other things going on in these earlier experiments – for instance, events caused by neutrons which could have mimicked those caused by a neutral current. And so, though neutral currents had been a matter of speculation for at least thirty years, experimenters didn't believe the evidence. They dismissed it as part of these background events.

In the 1970s new calculations and experiments were done by physicists who had in mind what the Salam–Weinberg theory told them they might find and some guidelines from the theory as to how to find it. As Weinberg tells it: 'One new thing in 1973 that was of special importance to experimentalists was a prediction that the strength of the neutral current forces had to lie in a certain range . . . This prediction provided a guide to the sensitivity that would be needed in an experimental search for these forces.'[10]

It's at this point that sceptics throw down a flag. They insist that Hanson's spectacles-behind-the-eyes were clearly at work, and that if you can't discover a physical phenomenon, one that's right there before your eyes, without the help of a theory to tell you it's there, then you have to wonder what other significant data you might be missing as a result of using this theory rather than another. Might not the data you're missing be the very data which would invalidate the theory? We do not have a situation in which we are scrutinizing independent evidence with complete detachment to find out whether a theory is correct. We have the theory leading us by the nose. Others insist, on the contrary, that taking science to task for not finding the neutral current before theory led the way is foolish nitpicking, that all this episode serves to illustrate is how much we need theory!

The story continues. 'What really made 1973 different', writes Weinberg, 'was that a theory had come along that had the kind of compelling quality, the internal consistency and rigidity, that made it reasonable for physicists to believe they would make more progress in their own scientific work by believing the theory to be true than by waiting for it to go away.'[11] But in 1976 there was a major set-back. Experiments at Oxford and Seattle, Washington, showed that neutral-current forces lacked some of the properties the electroweak theory predicted. Weinberg has this to say about the way he and other theorists reacted:

> Pierre Duhem and W. Van Quine pointed out long ago that a scientific theory can never be absolutely ruled out by experimental data because there is always some way of manipulating the theory or the auxiliary assumptions to create an agreement between theory and experiment. At some point one simply has to decide whether the elaborations that are needed to avoid conflict with experiment are just too ugly to believe. Indeed, after the Oxford–Seattle experiments many of us theorists went to work to try to find some little modification of the electroweak

theory that would explain why the neutral current forces did not have the expected kind of asymmetry between right and left . . . But nothing seemed to work.[12]

One problem was that the theory could not be altered satisfactorily and still agree with all the data that had previously supported it.

Then, in 1978, a new experiment at Stanford, California, was able to verify the predictions that the Oxford–Seattle experiments had called into question. Says Weinberg:

> Suddenly particle physicists everywhere jumped to the conclusion that the original version of the electroweak theory was correct after all. But notice that there were still two experiments that contradicted the theory's prediction for the neutral-current weak force between electrons and nuclei and only one that supported them . . . Why then as soon as that one experiment came along and found agreement with the electroweak theory did physicists generally agree that the theory must indeed be correct? One of the reasons surely was that we were all relieved that we were not going to have to deal with any of the unnatural variants of the original electroweak theory. The aesthetic criterion of naturalness was being used to help physicists weigh conflicting experimental data.[13]

I am reminded of a remark my son once made about the 'unscientific' way I count playing cards when I'm trying to find out whether there are fifty-two cards in the pack. He pointed out that if my first count indicates there are too few or too many cards, I recount. If the count comes out to fifty-two the next time around, I do *not* count a third time, I shuffle and deal. Perhaps that is not so unscientific! In most of the card games my family and I play, it soon becomes evident whether all the cards are there. Similarly, later experiments, though not repeating the Stanford experiment, have upheld the conclusion that the original version of the electroweak theory was correct.

A major influence of the electroweak theory on the course of physics was to encourage the development of accelerators powerful enough to produce the predicted weak force particles. This was an enormously costly, long-range undertaking, not the comparatively simple matter of a grant, a university budget, or even the budget of just one nation.

In 1983 physicists Carlo Rubbia, Simon van der Meer, and a team of 130 physicists at CERN in Switzerland finally carried out experiments designed to produce, if possible, three previously undiscovered particles predicted by Salam and Weinberg's theory. Now you might be thinking, 'Stop right there! One hundred and thirty physicists! That's certainly enough to meet the requirement that scientific evidence be corroborated and to ensure objectivity. Given the independence and eccentricity of physicists – certainly one of these would have removed the spectacles-behind-the-eyes for a moment and would have cried "The emperor has no clothes" if that had been called for!' However, there are others who would argue the contrary, that there is a point beyond which *more* observers tend to reinforce a point of view rather than ensure greater objectivity, and you don't necessarily pick the most rebellious physicist – the one who rejoices in being odd-man-out – when you're looking for a team player.

The anticipation and momentum generated by years of effort, millions spent, careers devoted to developing the equipment and designing the experiment – the point of view by now had been focused and strengthened by far more than the original scientific appeal of the theory. The world had bet heavily that the particles would appear. The men who had predicted them had already won a Nobel Prize (in 1979) for their theory. The discovery of the particles was really just the icing on the cake. In a situation like that, a lonely contradictory voice among 130 would be far less powerful, and less likely to speak at all, than when two or three scientists are working in relative obscurity – with all participants having a full understanding of the whole picture rather than each concentrating on a part of the experiment which requires his or her particular skill. No, the presence of 130 physicists at CERN is not the reason why we have a lot of faith in the electroweak theory, except insofar as their expertise was needed to run a reliable experiment.

The particles did indeed appear. First, in 1983, the two particles we call the W^+ and the W^-, and finally, in 1984, the Z^0, the messenger that carries the neutral-current interaction. They showed up with exactly the masses the electroweak theory predicted – another beautiful falling-into-place. How did they show up? No-one at CERN peered through a microscope and spotted the particles zipping around like tiny billiard balls. Physicists who designed the experiment judged that they would find them in debris from collisions of matter and antimatter moving at speeds approaching the speed of light. This debris appears as markings on photographic plates (see the illustration section). Rubbia and van der Meer and others among the experimental team scrutinized those for evidence of Ws and Zs among the debris.

'Direct experience of the universe'? It was that, surely. However, in such a situation, what exactly has been experienced does become a matter of interpretation, requiring great expertise and involving judgement calls; and the spectacles of theory are bound to influence that judgement. Someone has said that designing and interpreting a complicated physics experiment is so creative and subjective an activity that it resembles more a wine-taster blending a fine sherry than it does our naive picture of science at work. That may be overstated, but it's a metaphor that encourages us to recognize the subjective element in scientific discovery as an essential part of the process – and also to realize that we have been naive to think it could be otherwise. But it does not encourage us to debunk science. Regardless of how much creativity, subjectivity, and wearing of spectacles-behind-the-eyes goes into blending the sherry, the result either *is* a fine sherry or something that should be surreptitiously dumped into the nearest potted palm. The bottom line isn't really whether theory leads us, but whether it misleads us. Sometimes we know which it is only with hindsight. Scientists firmly believe that, in the end, we do find out.

Could we satisfy all scepticism once and for all by demanding that science remove all theoretical spectacles and start looking at the world as it really is? No. In fact, we could easily argue that the spectacles give us the vision to find truth we otherwise would never know was there. Without Salam and Weinberg's spectacles, would we have seen beyond the point of view provided by our niche in history, long after the electroweak symmetry was broken? We ask scientists to remember they're wearing spectacles-behind-the-eyes. We hope they'll try a few different prescriptions before they decide they're perceiving reality. But we can't expect them to wear no theoretical spectacles. We've already said that there is no possible way to escape viewing reality from one point of view or another. Think about my spy spectacles. One pair gave me one point of view. The second gave me another. But the view I had of the Seurat-like card when I was wearing neither was *also* a point of view. Who can say which of the three showed me 'reality'? Certainly none of them was allowing me to see everything there was to be seen. I want to switch metaphors to illustrate why theory is indispensable, and to take us still deeper into the problem of encountering independent reality.

Look at the room around you and imagine how you would have obeyed the instruction 'Draw a picture of this room and the things in it' when you were very young. Let's give you the benefit of the doubt and suppose you observed carefully and were fairly good at drawing. Each item would be in the picture: the door, the window, the books on the shelf, the lamp, the dog with the spot on one side, maybe even the mouse you knew lived behind the bookcase, which was visible only occasionally, and the two chairs and the desk. You could show the drawing to someone who hadn't been in the room, and they would know what things were in there.

Now imagine yourself when you were a little older. In the meanwhile, someone had taught you about perspective, and once again you drew a picture of the room – the same room, but with a difference. Now there was an additional order to the items, relationships between them showed clearly, and they probably threw shadows. Here's the question – Is the second picture (the one with perspective) a better representation of reality? What, if anything, have you lost in the second drawing?

There is a sense in which you have lost a bit of objectivity. That may sound unlikely, but in the drawing with perspective, what is perceived as reality in the room has become more dependent on you, the observer. Here is a room which (to misquote T. S. Eliot) has the look-of-a-room-that-is-looked-at – from one position and not from any other position – at one instant of time and not at another. In order to establish the relationships between the items in the picture and the pattern of light and shadows, you had to choose a point of view and give up all others. You the observer, the time of day you drew the room, your chosen position in the

57

room (though you can't be seen in the picture), and where the vanishing point is to establish the perspective, have become an essential part of the picture. If we extend this analogy and imagine looking at the room through a camera lens, with the possibility of changing the depth of the focus, the observer and the choices made by the observer become an even more confining point of view.

Now let's imagine that you are a cubist painter. Instead of choosing one point of view, one time of day, one vanishing point, one depth of focus, you paint many at once. Here on the left is a close-up of the mouse. Next to it are two rails from the chair back, here the side of the bookcase, here the head of the dog seen from the side, here from the front, here close, here far, here in shadow, here in light. And so on, and so forth. There are cubist paintings which incorporate so many different points of view and perspectives and details that they become more like an abstract rug design or a blur. A cubist painting, like the childhood drawing and the drawing with perspective, is a way of representing reality. In fact it is a much less biased representation of reality than your drawing with perspective. Its range of experience is far greater.

In science, the work of a naturalist or collector – the careful gathering and listing of data, the simplest 'observation' – is like the first drawing. With theory, we have the second drawing, the one that seems to give us a truer picture of the room. If in the second drawing one chair looks smaller than the other, when actually they measure the same dimensions exactly; if we can't discover that the dog has a spot on the side away from us and has two eyes, not one; that if we moved over a couple of feet we would see that there is a mouse behind the bookcase; that there is a desk behind us; well, that's a loss. Similarly, with a theory, for all we gain, we risk overlooking something significant that might make a different theory more useful to us.

What about the cubist painting? Would we prefer science to be like that – looking at reality from all points of view at once, wearing no spectacles, with all possibilities kept in mind? Some think that is exactly what science does. But you'll have to agree that whatever it may say about ourselves, our perceptions, and our thought patterns – and whatever risk we may run of distorting reality and fooling ourselves – the perspective drawing is almost certainly more useful than the cubist drawing in helping us find our way in the room. In science, we've discovered that somewhere in the range between the childhood drawing and the cubist drawing we have to agree on a point of view, or a limited number of points of view, that are useful and meaningful to us. We don't claim to have chosen the ultimately correct and complete image of reality. A room can be looked at not just from hundreds but from an infinite number of perspectives, and Poincaré insisted that 'if a phenomenon admits of a complete mechanical explanation, it will admit of an infinity of other [mechanical explanations] which account equally well for all the peculiarities disclosed by experiment.[14] Do you see our problem – now that you are wearing the spectacles which *I* have provided?

How then do we choose our view of the room?

Poincaré tells us that when one explanation reveals relationships that the other hides from us 'we may regard it as physically more true than the other, because it has a richer content.'[15] And although several theories might be equally plausible explanations of the data, each will make different additional predictions that can be tested. We've already discussed some of the characteristics that cause one theory rather than another to make it into the textbooks and graduate curricula and plot the course of science.

But one of the most powerful criteria is one we would least expect.

THE MUSE OF SCIENCE: IS TRUTH BEAUTIFUL?

Among the many tales about the great mathematical physicist Paul Dirac is one told by his friend and colleague Jagdish Mehra. Dirac and Mehra met for the first time dining at high table in St John's College, Cambridge. Mehra was nervous about being seated beside the legendary Dirac. 'The weather outside was very bad,' he recalls, 'and since in England it is always quite respectable to start a conversation with the weather, I said to Dirac, "It is very windy, Professor." He said nothing at all, and a few seconds later he got up and left the table. I was mortified, as I thought that I had somehow offended him. He went to the door, opened it, looked out, came back, sat down, and said, "Yes".'[16]

Here, surely, was a man who lived by the scientific method! A man who, like Darwin, 'insisted on the clearly ascertained report of the senses'.

Consider then the following quotation from Dirac himself: 'It is more important to have beauty in one's equations than to have them fit experiment . . . because the discrepancy may be due to minor features which are not properly taken into account and which will get cleared up with further developments of the theory . . . It seems that if one is working from the point of view of getting beauty in one's equations, and if one has a really sound instinct, one is on a sure line of success.'[17] Beauty is a subjective matter – 'in the eye of the beholder', we are told – what could be more subjective than that? But beauty is a familiar pointer in physics.

Nearly all physicists understand well what Dirac meant when he spoke about beauty. Mathematician G. H. Hardy wrote: 'The mathematician's patterns, like the painter's or the poet's, must be beautiful. The ideas, like the colours or the words, must fit together in a harmonious way. Beauty is the first test.'[18] Weinberg said: 'I believe that the general acceptance of general relativity was due in large part . . . to its beauty.'[19] Physicist Murray Gell-Mann calls it one of the great mysteries: 'Why is our sense of beauty and elegance such a useful tool in deciding whether a thing is good or bad?'[20] John Wheeler says God or evolution has formed the minds of some of us in such a way that our instinctive ability to recognize beauty is a practical tool for finding truth. Perhaps recognizing when you're onto something and calling that 'beauty' has become instinctive as successive generations of physics students have watched what works among their elders and in examples in their textbooks, and have become conditioned to pick up subtle clues. Whatever the explanation, most physicists would agree that it is in large part Dirac's 'really sound instinct' for getting on the track of beauty that makes a great physics theorist.

Dirac's words, 'It is more important to have beauty in one's equations than to have them fit experiment', are not a complete abandonment of objectivity. What physicists mean by beauty isn't exactly what the rest of the world means by beauty, though it's close. Most of us have experienced beauty in the sense of everything working out in a soul-satisfying, harmonious manner.

That's certainly a part of what moves us in art, music, poetry, nature, and even in a beautiful face or body, the coming together of many disparate elements in a way that seems inevitable, effortless, intensely pleasing, beyond our expectations.

Beauty in physics similarly has to do with a falling-into-place that appears little short of miraculous. It implies simplicity, elegance, and mathematical consistency and creativity. These are the qualities that make such theories as superstring theory and wormhole theory appealing and convincing, not the fact that anyone has ever observed a superstring or a wormhole or that anyone hopes to in the foreseeable future. Even if a theory clashes with experiment and observation, as the electroweak theory did for a time, it isn't illogical to think that it may be the experimental results which are misleading rather than the theory.

Mathematical consistency is punishingly difficult to achieve, and that's one of the reasons why it's convincing when it is achieved. There is so much well-established maths and physics with which a theory must be consistent. It's a little like doing a crossword puzzle in which clues and words familiar to you converge in such a way as to form a word you've never heard. Even before consulting the dictionary, you're certain your unknown word must be correct. If it weren't, you would be obliged to reconsider large parts of the puzzle that you know you have right because they themselves fit together so intricately and successfully. If your word isn't in the dictionary, you begin to wonder about the dictionary.

The instinct for beauty has played a part in the history that led to the acceptance of each of our greatest twentieth-century theories. However, beauty even in the form of rationality, mathematical consistency, and a seemingly inevitable falling-into-place is in the end considered less conclusive than direct experimental testing and observation. Black holes are mathematically beautiful, but we still wanted like anything to find one. For many like Dirac, beauty has been an exceptionally reliable pointer. However, this pointer is not infallible. Wasn't it this same instinct for beauty in the sense of rationality and intelligibility that told Einstein, Louis de Broglie, and Erwin Schrödinger (founding fathers of quantum physics) that events on the quantum level could *not* be inherently uncertain? And weren't they wrong? . . . or so we think – so far.

It's also true that scientists sometimes have no choice but to explore unbeautiful territory, where everything seems out of order and wrong. An unbeautiful problem which still exists is an apparent contradiction between general relativity and quantum theory which you'll be reading more about in Chapter 5. These two outstanding theories of our century serve us magnificently on both the theoretical and the practical level. They underlie much of our modern technology. Nevertheless, if both are true, we are left thinking that the universe ought either to be curled up into a small ball or to have expanded in such a way that galaxies couldn't form. A glance around us tells us that neither is the case. This sort of inelegance is unsettling, in spite of the overwhelming success of the theories involved.

We follow the pointers of beauty and mathematical logic from theory to theory, deeper into the mysteries of the universe, with the hope that if we follow it far enough we'll come eventually to an idea behind everything, whose beauty will far surpass any we've encountered before. Scientists such as John Wheeler feel certain that this will not be a complicated idea. It will be simplicity itself. In the yearning to find this simple, beautiful idea, the search for knowledge in physics becomes intermingled with the search for God. 'Sing God a simple song,' wrote Leonard Bernstein in the opening solo of his *Mass*, 'for God is the simplest of all.' Find a word for God which implies ultimate truth without insisting on the notion of a person, and many an agnostic scientist will sing along with Bernstein.

Our faith in mathematics and logic leads us to believe that if a thing isn't mathematically and logically consistent it can't be true – it isn't allowed to be true. That's a tight constraint upon what can and cannot exist or take place. If mathematical consistency becomes more and more difficult to achieve as we approach ultimate truth, as theorists insist it does, then it may be that difficulty which finally narrows us down to one mathematically consistent equation underlying the entire universe. Is it possible that there was only one way God could have made the universe without violating mathematical consistency – which isn't allowed? That's a question which intrigued Einstein: 'What I'm really interested in is whether God could have made the world in a different way; that is, whether the necessity of logical simplicity leaves any freedom at all!'[21] If God is constrained by mathematical consistency, then mathematical consistency is stronger than God – or even IS God. And where did mathematical consistency come from?

It's a question of profound importance whether mathematical consistency required an Inventor. I've heard it asked at the end of public lectures on physics: 'Is mathematical consistency as we know it the only way it COULD be – or is it conceivable it could be something different? Did anybody have to choose that it be the way it is?' If the lecturer is a scientist or mathematician, he or she may answer that mathematical consistency just IS. In the words of G. H Hardy, '317 is a prime number, not because we think it so, or because our minds are shaped in one way rather than another, but *because it is so*, because mathematical reality is built that way'.[22] If there isn't any other way it could be, we don't need to find a cause for it, much less a creator who decided that it be so. The uncaused, unexplainable 'First Cause' of the universe may be mathematical consistency. Full stop.

A few members of the lecture audience will find the statement that mathematical consistency 'just IS' unsatisfactory, not necessarily for religious reasons. The scientist or mathematician may think these hold-outs naive, but it could be argued that they are capable of more divergent thinking than the scientist/mathematician – that they are less captive to a point of view. There is a gut feeling abroad that anything COULD be. Is it so inconceivable that a reality COULD exist in which 317 is not a prime number? Our not being able to imagine it proves nothing at all. Hardy comes near conceding the point by saying that 'mathematical reality is *built* that way.' But he almost certainly didn't intend to concede the point. Maybe he, like the rest of us, simply lacked the vocabulary to describe uncaused fundamental truth. Whether or not mathematical consistency must be what it is and whether it required an inventor – unfortunately, these are questions mathematics can't answer conclusively about itself. There are other such questions.

DOES TRUTH SURPASS PROOF?

Mathematics seems to most of us a sterling example of clarity and objectivity. If there is a universal language, in the most literal sense of the term 'universal', maths is it. Early in life we learn about a direct correlation between maths and reality. Put two apples in a box, put in two more, and there are four apples in the box, just as maths infallibly predicts. From such simple beginnings we grow to have faith that this correlation with reality will hold in situations far beyond anyone's ability to demonstrate with picturable objects: $10^{40} \times 10^{10} = 10^{50}$. We believe that, though we will never see it happen with apples. We assume all this holds true in the Andromeda galaxy and at Quasar P.C. 1158 + 4635 as surely as in our own back yard. We agree with Galileo that 'The Book of Nature is written in mathematical characters',[23] hardly ever stopping to realize how amazing and even unlikely it is that nature should always seem so faithfully to bear out our mathematical predictions. We've already seen that for those who become mathematicians and physicists, faith in mathematics may grow so strong that for them mathematical consistency will rival experimental results and observation as convincing evidence of truth. For some it seems a stronger concept than the concept of God.

Isn't mathematics our one sure, unclouded window to reality? Why should we not consider it even stronger than observational and experimental evidence, especially having seen how ambiguous the notion of a direct encounter with independent reality can be? British scientist Jonathan Powers has expressed it well: 'We experience mathematics as a source of Absolute Authority and as a repository of Absolute Truth, uncompromised by mere human interest. Mathematical proofs are implacable, and cannot be deflected by bluff or bargaining.'[24]

Interestingly enough, one of the things that maths *can* prove is that there is truth beyond maths' ability to prove. It was Kurt Gödel, a German mathematician, who alerted us to this problem. In 1931 he came up with a theorem now known as Gödel's Incompleteness Theorem. Gödel showed that in any mathematical system complex enough to include the addition and multiplication of whole numbers, there are propositions which can be stated – that we can even *see* are true – but which cannot be proved or disproved mathematically within the system. It wouldn't matter so much if these unprovable (and un-disprovable) propositions were just peculiar oddities on the fringe of mathematics, but they are not. They include extremely significant results. The addition and multiplication of whole numbers is certainly not exotic territory!

In Chapter 2 we saw that the assumptions underlying the scientific method are not capable of being proved or disproved by the scientific method. If Gödel was right, belief in mathematics also requires a leap of faith. All significant mathematical systems are open and incomplete. Even in mathematics, truth goes beyond our ability to prove that it is true. One definition of religion has it that a religion is a system of thought which requires one to believe in 'truths' which can't be proved. If that's what a religion is, then according to Gödel's Theorem, mathematics is a religion. In fact, mathematician F. De Sua has remarked that it seems to be the only religion that has proved it *is* a religion.[25]

The implications of Gödel's discovery are far-reaching and disturbing. As John Barrow points out in his book *Pi in the Sky*,[26] we have learned that it isn't possible to prove for certain that any system rich enough to include addition and multiplication of whole numbers is self-consistent. That means that systems such as geometry, arithmetic, logic – any of the mathematical systems which physicists rely on – may turn out to be internally contradictory, and there is no way we can ever prove they are not. What can we conclude, if the maths worked out one way leads to one prediction, and worked out another way leads to a contradictory conclusion? In fact, we can conclude that we are lucky to have discovered the contradiction! There isn't any sure way to discover such contradictions methodically. Discovering them is largely a matter of chance.

The great speculative theories dealing with the origins of the universe and the unification of the forces rely very heavily on mathematical consistency. Since we cannot ever be certain where in our mathematics contradictions may lurk unsuspected, our situation is precarious. One theory builds on another. We can't escape the suspicion that we may be constructing a very ephemeral house of cards. On the other hand, we saw in Chapter 2 my daughter's childhood impression that nature has not made use of all the possibilities of geometry. Perhaps nature has not made use of all the possibilities of arithmetic, and has avoided the pitfalls. Barrow points out that 'it may be the case that physical reality, even if it is ultimately mathematical, does not make use of the whole of arithmetic and so could be complete', even if mathematics is not.[27] Unfortunately for theoretical physics, we can't assume that is so.

We've said that we meet obstacles in the search for ultimate truth that we don't meet in the search for partial truths and predictability. One of the big questions is, how will we know when we've found ultimate truth? How can we prove we've discovered it? Gödel's Theorem with all its implications for the concept of mathematical consistency – found as it is in the area of human knowledge that has traditionally seemed most able to establish incontrovertible proof – is discouraging.

THE ELITE OF SCIENCE

We said early in this chapter that point of view in science comes not only from scientific technique or theory but also from sources we think of as less legitimate influences. Here are examples:

When we try to form a picture of what the truth is about the universe or a part of the universe, we bring to the task preconceived notions so deeply ingrained that we are barely conscious of them. Such notions as 'What I would *prefer* the truth to be' (I am uncomfortable with any other version); 'What I think the truth *ought* to be' (If I had created the universe, I would certainly have done it thus); 'What I imagine it is' (I'm guessing now); 'What I fear it is' (I don't even want to consider that); 'What it simply must be' (Anything else would be unthinkable) are certainly not objective. These are personal points of view – just the sort the scientific method is supposed to weed out when they are unsupported by data. The cherished personal notions of eminent scientists – 'icons of the scientific quest', as Powers calls them – make for difficult and slow weeding. The opinions and prejudices of the acknowledged experts of our generation strongly influence which theories other scientists take seriously, which they scoff at, and what avenues of inquiry they follow.

Scientists are particularly loath to relinquish the last form of prejudice in the list above – 'It must be true because anything else would be unthinkable.' As the Nobel physicist Subrahmanyan Chandrasekhar said of Arthur Eddington: 'He was a great man. He said that there must be a law of nature to prevent a star from becoming a black hole. Why should he say that? Just because he thought it was bad? Why does he assume that he has a way of deciding what the laws of nature should be? Similarly, this oft-quoted statement of Einstein's disapproving of the quantum theory: "God does not play dice." How does he know?[28] Chandrasekhar was referring to Einstein's belief, which Einstein never relinquished, that the quantum level could not be inherently uncertain.

Chandrasekhar was in a position to appreciate the significance of Eddington's prejudice.[29] In January 1935, when he was a twenty-four-year-old student from India at Cambridge, he presented a paper at a meeting of the Royal Astronomical Society. The paper announced a significant mathematical discovery he had made about what happens when a massive star runs out of fuel and can no longer support itself against the pull of its own gravity. Chandrasekhar had every reason to expect this presentation to be the making of his career. He'd been encouraged by the interest Sir Arthur Eddington had taken in what he was discovering. Eddington was the senior statesman among physicists at Cambridge at that time. Today we call the dividing line that Chandrasekhar laid down in that paper – the dividing line between the mass of stars which might stop collapsing and go on existing as White Dwarf stars and those which would continue to collapse – the Chandrasekhar limit. What Chandrasekhar had discovered was to become a vital part of black hole theory, and it was partly on the basis of this early work that he received the Nobel Prize in 1983.

But no-one predicted this on 11 January 1935.

Chandrasekhar didn't say in his presentation that day that a massive star would become a black hole. He couldn't have used that term in any case, since John Wheeler wouldn't invent it until thirty years later, but he was cautious enough to avoid saying how the collapsing star would end up, leaving that tantalizing question open.

Eddington, who spoke next, was not so reticent about the implications of Chandrasekhar's findings. 'I suppose,' Eddington said, '[the star] gets down to a few kilometres radius, when gravity becomes strong enough to hold in the radiation, and the star can at last find peace.' He called that result unthinkable, the perfect example of a *reductio ad absurdum*. 'I think there should be a law of nature to prevent a star from behaving in this absurd way.'[30] Such was Eddington's influence in physics that his audience readily joined him in his scorn, even though no-one, including Eddington, could fault Chandrasekhar's logic or calculations. The situation Eddington said couldn't possibly exist – while Royal Society members laughed with him – we now of course do call a black hole. Chandrasekhar remembers standing that night in the deserted common room back in Cambridge, thinking 'This is how the world ends, not with a bang but with a whimper.'

Today, Chandrasekhar doesn't regard that incident as the crushing blow it seemed to him then, although the continuing disagreement with Eddington ruled him out of any tenured position in England and finally caused him to move to other subject areas and not return to black holes for many years. In spite of initial doubts, he did continue his career, in America not in England, and he thinks that he benefited as a scientist and a person from not having success come so early.

But what about the impact on physics? 'Suppose Eddington had decided that there were black holes in nature . . . It's very difficult to speculate,' Chandrasekhar says. 'Eddington would have made the whole area a very spectacular one to investigate, and many of the properties of black holes might have been discovered twenty or thirty years ahead of time. I can easily imagine that theoretical astronomy would have been very different. It's not for me to judge whether that difference . . . well, the difference would have been salutary for astronomy, I think I would say that.'[31]

Though we hear the epithet 'tyranny of old men' used to describe the elite of science, this elite is by no means made up only of those who are over the hill and no longer capable of doing meaningful work. Nor are the elite only the 'icons' like Eddington. They are also all those on university committees, government committees, grant committees, editorial boards, and corporate boards, who determine whose theories and proposals are taken seriously, whose paper gets published, whose theory is tested. As someone has described the competition for promotions, grants, laboratory space, and telescope time, 'There are too many at the trough and the swill is thin.' Politics, economics, and fads within the scientific community have a significant impact on what theory becomes a 'spectacular one to investigate', or even a *possible* one to investigate – and what as a result emerges as scientific knowledge.

Another consideration is that the mentor system is still very much alive in science. If you are a physics graduate student, after finishing a course curriculum designed to give you a broad view of the field, you narrow your area of study and do graduate research under the tutelage of an individual or group of individuals whose current work most interests you and where there is grant money available to keep you alive – if you're lucky, with one of the 'icons'. You align yourself at least for a time with the mind-set and theoretical views of your mentors, and your own research agenda is related to theirs, often contributes to theirs. That doesn't necessarily mean that your work isn't original, and that there isn't, along with some hero worship and genuine admiration, a certain amount of grumbling about this tyranny. However, at least as far as Ph.D. and postdoctoral research are concerned, you're unlikely to stage a serious rebellion. These people are going to decide whether you get your degree, and their connections will, you hope, land you a good job. By the time you are free to go your own way, your particular slant on science and what, through your efforts, will emerge as scientific knowledge – unless you really are unusually independent – are likely to be coloured by the views of your former mentors, especially if their work is prestigious.

THE SPIRIT OF THE TIMES

The mind-set of our culture and our historical setting also play a strong role in determining what theoretical proposals are taken seriously and what emerges as scientific knowledge. This happens partly as the result of pressure coming from outside the scientific community, but it's also the result of the fact that scientists themselves are part of this culture, and they, like everybody else, have absorbed its fashions, its values, and its standards of morality.

The scientific method itself is not a method designed to make moral judgements or value judgements. In principle a scientific decision about whether something is true is not a decision about whether it is good. Truth may very well offend us, cause us problems, raise acute moral dilemmas, not be 'politically correct'. Truth may not seem at all 'good' by current standards of what is good, though the same truth might have seemed good a century ago or in another culture. Why then should anyone say that what suits our current standards is more likely to emerge as 'truth'?

Standards have a way of translating to priorities. Society and groups within society force science to concentrate on finding what we most want to find. We encourage the testing of theories we like and where the findings promise to be advantageous to us, and not those where the results might be of little use or repugnant to us. Much of this encouragement and discouragement comes from outside the scientific community, and it is particularly strong when the cost of testing a theory requires a large public or private outlay of funds.

We demand cures for cancer or Alzheimer's, not those diseases which afflict less wealthy countries (who couldn't pay for the pharmaceuticals) or only a tiny portion of the population – until they impinge on us, as AIDS now does. An item in *Newsweek* in November 1992 stated:

> Drug companies have largely given up looking for [remedies for malaria]. From a commercial perspective, it makes little sense to turn out costly pharmaceuticals for people who can't afford shoes . . . Altruism has never played a big role in malarial research. Quinine enabled Europe to colonize the tropics. Chloroquine grew out of efforts to protect U.S. troops abroad. Without an empire or an army on the line, the developed world will need a new rationale for fighting malaria. At the moment, the best one is that 2.1 billion people – about 40 percent of the world's population – are in danger.[32]

We favour science which boosts national prestige and security: we want to win the race into space; we want effective weaponry and missile defence systems.

We give preference to what suits our economic aspirations: give us the inventions that will enhance competitiveness – innovative goods and services that will strengthen our industry for the international trade conflicts of the 1990s and later.

70

We jump on bandwagons: for a time the rage is 'pushing back the frontiers' – space exploration and super-colliders, study of DNA that might allow us to clone dinosaurs – science according to Steven Spielberg and George Lucas. Then, none of that – give us useful science; of what possible practical value would the discovery of the Higgs particle be?

We raise issues which make it difficult for some types of research to take place: don't experiment with animals, humans, or human foetuses; don't experiment with genetic engineering and cloning.

The statement of Charles Darwin's grandfather Erasmus Darwin in 1783 upon founding the Derby Philosophical Society that the society would seek 'gentlemanly facts' sounds quaint to us today, but we are not so far from that when we seek 'politically correct facts'. We make up our minds regarding issues on the basis of values and principles which are important to us and assume that research can do no other than show that we are right, not contradict us: show us for instance that 'intelligence' is largely a product of environment, not inheritance, so that inequalities are something we can correct socially. With regard to some loaded issues, such as those having to do with the roots of sexual preference, or genetically or racially linked IQ, it is extremely difficult for scientists to get funding for research where there is even a risk of finding what we don't want to know. Many of us will not accept unwelcome answers as valid findings. We can argue this is a good thing: life is unfair enough without looking for scientific excuses for even more unfairness.

But all of this leads us to a question: Is any system of values or principles a stronger concept than scientific truth? It seems that for most of us the answer to that question may be yes. In spite of our enormous faith in science, when scientific evidence conflicts with our most deeply held principles and beliefs, we do not easily capitulate. Einstein, you may remember, was unwilling to accept that the quantum universe was inherently uncertain. He felt that quantum uncertainty must be the result of limitations in our measuring ability. In order to show that he and those who agreed with him – the minority among physicists – were right, he proposed an experiment. It wasn't until after his death that the difficult obstacles involved in carrying out the experiment were overcome, and French physicist Alain Aspect, CERN physicist

John Bell, and colleagues were able to carry it out and interpret it. The result – Einstein was wrong. The quantum world *is* inherently uncertain. In his book *Superforce*, British physicist and author Paul C. W. Davies writes the following:

> Several months after Aspect published the results of his experiment I had the privilege of making a BBC radio documentary programme about the conceptual paradoxes of quantum physics. The contributors included Aspect himself, John Bell, David Bohm, John Wheeler, John Taylor, and Sir Rudolph Peierls. I asked all of them what they made of Aspect's results and whether they thought that common-sense reality was now dead. The variety of answers was astonishing.
>
> One or two of the contributors felt no surprise. Their faith in the official view of the quantum theory as enunciated long ago by Bohr was so strong that they felt the Aspect experiment merely provided confirmation (albeit welcome confirmation) of what was never seriously in doubt. On the other hand, some were not prepared to leave it at that. Their belief in common-sense reality – the objective reality sought by Einstein – remained unshaken. What would have to go, they argued, was the assumption that signals could not travel faster than light. There must be some 'ghostly action at a distance' after all.[33]

Dutch scientist A. van den Beukel has commented: 'You rub your eyes in amazement . . . Those who hold the one belief do not need any confirmation; they knew it all along. Those who hold the disputed belief are unshocked by what seems to be an overwhelming argument for the prosecution and are ready to throw overboard one of the most fundamental principles of the whole of science, as if it were nothing.'[34] We don't know, of course, what Einstein's reaction would have been to the outcome of his proposed experiment.

We've seen that no experimental results need be considered absolutely conclusive. We are not necessarily being 'unscientific' when we refuse to accept as 'the final word' a result which seems to undermine a principle in which we believe strongly. That said, we must admit that ideology, scientific and other, can play a significant role when we judge how conclusive an experiment *is*. Regarding common sense reality on the quantum level of the universe, within our society as a whole we don't have an ideological spirit of the times ruling strongly either for or against – certainly not strongly enough to alter the course of science. When it comes to social issues such as some we spoke of earlier, we can't state that so confidently.

We're all to a certain extent prisoners of the mind-set of our culture and time in ways so inherently part of us that none of us can discern exactly how we and our science are influenced. It's easier to see biases in other cultures and historical eras than our own, but we can't look thoughtfully at human history and come away believing that our own culture is for some reason the exception – free of biases that affect our perception of the world.

THE ESSENTIAL GODLESSNESS OF SCIENCE

While we're on the subject of values and principles: Do religious views affect what emerges as scientific knowledge?

We would be surprised today to find, except among scientific creationists, any scientist openly allowing the Bible or religious teaching jurisdiction over what is true or false among scientific findings. But what about more subtle influences? And what about the argument that some scientists allow their atheism to dictate what they accept among scientific findings and what theories they prefer?

The assumption that science is a Godless pit of atheists is false. Many scientists do believe devoutly in God and many others are agnostics, not atheists. However, if someone put forward a theory predicting that prayer for healing improved chances of healing, and proposed to put that to the test, hardly anyone in the scientific world – atheist, agnostic, Jew, Christian, Moslem, Buddhist, or anything else – would treat that as serious science. Why? Why should it seem essential that we leave God totally out of science in order to do valid science? To be completely accurate, we must mention that serious Jews and Christians have reason not to take such a theory and testing seriously on the grounds that one of the primary tenets of their religion is 'Thou shalt not put the Lord thy God to the test.'[35] But that can't be the only reason why science avoids such questions.

73

One of the underlying assumptions of science is that knowledge about the universe is accessible to us. The way we seek to uphold that assumption is by trying to break down any barriers that threaten to make any areas of knowledge inaccessible, hoping that whatever appears inaccessible and beyond our understanding will eventually become an open book. Evidence from quantum theory and chaos/complexity theory argues otherwise. But faith in the accessibility of the universe still remains a foundation stone of science.

If there is a God, that almost certainly means everything is not accessible and understandable by human discovery and reason alone. There is knowledge we'll never have unless God himself chooses to reveal it. Scientists who believe in God say that their knowledge of God enhances their science, but even most of these believers still practise science on the premise that the unknown is entirely fair game, not forbidden territory. The more acceptable, scientific way of studying an event which might be attributable to divine action would be to try to find an explanation that is not beyond human understanding, as Weinberg puts it, 'to assume that there is no divine intervention and to see how far one can get with this assumption.'[36] If all this sounds like a prejudiced point of view, it arguably is – but it doesn't necessarily lead to prejudiced scientific results. It is not at all unreasonable to think that if there is a God, he would be better served by trying to falsify him, and failing, than by trying to prove he exists. And God *can*, we suppose, look out for himself!

We have a good example in twentieth-century science of the paradoxical way attempts to falsify a theory help verify it. A controversy raged for fifty years, first over whether the universe is expanding and later (when it was inescapably clear that it is expanding) over whether this expansion means the universe had a beginning, the Big Bang. Support for the notion of a creator in the Big Bang theory arises not only from the fact that the theory provides a moment when creation *could* have occurred (as an eternal universe does not) but also from the fact that at the beginning of the universe we encounter the unexplainable. Laws of physics as we know them break down; there is an initial event the cause for which is unknowable.

As American astronomer Robert Jastrow wrote in his book *God and the Astronomers*,

I am an agnostic in religious matters. However, I am fascinated by some strange developments going on in astronomy – partly because of their religious implications and partly because of the peculiar reactions of my colleagues . . . Theologians generally are delighted with the proof that the Universe had a beginning, but astronomers are curiously upset. Their reactions provide an interesting demonstration of the response of the scientific mind – supposedly a very objective mind – when evidence uncovered by science itself leads to a conflict with the articles of faith in our profession. It turns out that the scientist behaves the way the rest of us do when our beliefs are in conflict with the evidence. We become irritated, we pretend the conflict does not exist, or we paper it over with meaningless phrases.[37]

When Jastrow refers to 'beliefs', he doesn't mean only religious beliefs, but quite clearly the reactions of many, beginning with Einstein, to the idea that the universe was expanding, were far from cool-headed and objective. 'This circumstance [of an expanding universe] irritates me,' Einstein wrote.[38] Allan Sandage, whose work was important in confirming the expansion theory, nevertheless said, 'It is such a strange conclusion . . . it cannot really be true.'[39] When it became evident that the universe, regardless of anyone's preference, was indeed expanding, Hermann Bondi, Tom Gold, and Fred Hoyle came up with 'Steady State theory', a measure to explain the expansion of the universe in a way that would not require the universe to have had a beginning. The three of them were outspokenly resistant to an explanation which seemed to support a biblical view of creation, and they were not alone in their disappointment when observational evidence supported the Big Bang rather than Steady State theory. For reasons entirely apart from scientific objectivity, the Big Bang pill was a bitter one to swallow, and a few still have it hiding behind a tooth.

That history of opposition to the Big Bang theory is, however, one of the reasons it is so convincing today. When a theory has to fight its way against scepticism and opposition within the scientific community and when there is a serious competing theory, it is far more likely to satisfy Popper's requirement that it be tested for as many as possible of the predictions that would disprove it, and the evidence favouring it must be extremely convincing. As was not the case with the electroweak theory, a large segment of the physics community was reluctantly won over to the Big Bang theory by data they didn't much care to find.

However, as we will see in Chapter 4, physicists did not give up hope of getting past the slammed door that we thought we had encountered in the Big Bang singularity. Some of these are at least pleased to show as a by-product of their theory that we don't need God after all. Hawking makes a very big point that his no-boundary proposal shows how the universe could just BE ('What place, then, for a creator?'[40]).

In a similar vein, Richard Dawkins states that one of his primary purposes in writing his bestseller about evolution, *The Blind Watchmaker*, was to show 'evolution as the true explanation for the phenomena that Paley thought proved the existence of a divine watchmaker.'[41] Many have asked why, for either Hawking or Dawkins, it was necessary to bring the other suggestion – God – into the discussion at all? Why not just stick to the science?

A Brief History of Time and *The Blind Watchmaker* are two of the finest books ever written for the popular science audience, and both authors seem obsessed with God. Whether or not it is true, both give the impression that the fact that the scientific theory they are writing about erases our need for a God is far more reason for celebration than the fact that the theory makes a new part of this mysterious universe accessible to human beings. This can't be called a religiously neutral point of view. Science, for Hawking and Dawkins, is *not* essentially Godless.

Anything that influences our choice of which theory will lead the way is a potential influence on the future course of science and upon what will emerge as scientific knowledge. Clearly, we no longer do science, in the phrase coined at the Cavendish Labs in Cambridge, 'with sealing wax and string'. It's a trend in the late twentieth century, especially in physics, for theory to run not just a little way but very far ahead of experiment, and for it to influence decisions as to who will get the money for decades in the future to buy the expensive modern equivalents of the sealing wax and string. Theory plots the course of science, and it isn't only scientific thinking that determines which theory will lead us. The choice is a complex and haphazard affair. Everything from pure aesthetics to the lust for power, fame, and wealth plays a part.

Should we agree with that minority who say scientific knowledge is what it is almost entirely as a result of forces that have nothing to do with any encounter with reality? The recognition of all the forces involved need not leave us so pessimistic as that. Weinberg writes:

> A party of mountain climbers may argue over the best path to the peak, and these arguments may be conditioned by the history and social structure of the expedition, but in the end either they find a good path to the peak or they do not, and when they get there they know it . . . I cannot prove that science is like this, but everything in my experience as a scientist convinces me that it is. The 'negotiations' over changes in scientific theory go on and on, with scientists changing their minds again and again in response to calculations and experiments, until finally one view or another bears an unmistakable mark of objective success. It certainly feels to me that we are discovering something real in physics, something that is what it is without any regard to the social or historical conditions that allowed us to discover it.[42]

Is Weinberg right? Will science, as an instrument for learning about reality, prove strong enough to overcome all the glitches and stumbling blocks, the fads and false leads, the good but mistaken intentions, the arrogance and the assumptions, the din of many voices urging us down one path or another? Is it correct to believe that eventually the truth will out? 'Muddling to discovery' was what American physicist W. Peter Trower called it.[43] That sounds more like Monty Python's knights than the Bayreuth laser-lit gods and heroes. Perhaps it is the intrinsically human way of getting there. It's also human to wonder about the roads dimly perceived but never followed, the 'hints' not even 'half-guessed', and the points of view that may lie completely beyond human conception.

It's time to ask how much the scientific method itself gives us a biased view of reality.

AT THE LIMITS OF SCIENTIFIC TRUTH

Religion, philosophy, art, music, poetry, literature – none of these instruments probes the world as confidently and systematically as science. Someone has said that they caress and butcher the world, while science performs laser surgery. Nevertheless, the arts and humanities have stretched the boundaries of human experience and given us insights and explanations that have the unmistakable feel of truth about them. As science is unable to do, they incorporate and even celebrate the unexplainable, the freak, the uncategorizable, the unpredictable, the senseless, the unique, the miraculous, the absurd and the irrational. The traditional study of science, with its penchant for predictability, rationality, elegance, and simplicity, seems by comparison an escape to a formalized, artificial world.

Nevertheless, if there is such a thing as objective truth, it must be the same for the artist, the philosopher, the religious, the poet, and for the scientist. How can this be? One of the finest teachers I ever had, who had lived an adventurous life that had taken him all over the world into many cultures, insisted as an old man that he couldn't stomach the modern rhetoric about all human beings being alike. 'Don't you see,' he asked us, 'that the wonder of it . . . the glory of it . . . is how *different* we are!' Science, religion, art, literature, and music all study the same reality. The wonder of it – perhaps the glory of it – and certainly the confusion of it – is how differently they see it.

It's easier to recognize the biases of other generations than it is to recognize our own. At present we're able to see that the point of view dominating science from the time of Newton until well into our own century, the point of view which saw reality in terms of predictable systems, was a distorted and limiting point of view. Predictable systems lent themselves more readily to scientific study which ends in meaningful and helpful results, and for that reason anyone choosing a line of inquiry was likely to choose an area which showed promise of systematization. Over the years, what began as a hope became a definite impression, that everything in the universe, all the complication and variety that is obviously there, would eventually resolve to predictable systems. Today chaos and complexity theories show that predictable systems are the exceptions, not the norm. Science has been operating from a biased point of view, and finding what she expected to find.

Science, like other areas of human knowledge, evolves, and there is no reason to think that all blinkers science may be wearing today will be there tomorrow. Nevertheless, there is a broader and perhaps more permanent sense in which the scientific method may be a limiting point of view:

Suppose we find intelligent beings on another planet. How different might 'reality' look from the point of view of *their* science? It is one of the assumptions of science that there are fundamental laws which hold for the entire universe. But these laws would not manifest themselves in precisely the same ways on other planets. We know that though the gravitational constant is not different on the moon, the experience of gravity on the moon is different from on the earth. Beyond such easily explainable discrepancies, are there other more fundamental differences we might find in an alien science?

On the individual level of perception, our minds to a certain extent invent our views of a chair based on previous experience of such objects and previous experience of 'seeing' – assumptions about size, distance, and perspective we learned in infancy. Psychologists tell us that it is almost impossible for a person to describe an object which can't be linked in any way with anything he or she has experienced before. If a person blind from birth is given sight, that person doesn't immediately know how to see. Even if beings from another planet have five senses similar to ours, would they see anything like *my* chair?

We suspect that thought processes which developed in response to problems of survival dictated 'how we look' long before there ever was a thing called science. We know that science evolves partly in response to problems society feels it needs to solve. The same will have been true for the development of perception and scientific knowledge on another planet. Have our methods of discovery and reasoning about the universe turned out at all similar? Would even our logic seem logical to alien minds? *Is* there only one possible method that leads from individual mind's-eye views of the chair toward the chair-as-it-is-in-itself? If not, have we humans found the best method? And what does our method not allow us to discover?

FIRST STEPS BEYOND THE MIND'S-EYE VIEW

At the most basic level, our scientific method takes us beyond all our unique starting points and allows us to expand upon our individual, private mind's-eye pictures of reality by showing us relationships. We saw in Chapter 2 that just because I say the chair is brown and you also say the chair is brown doesn't mean we're seeing the same colour. Brown is just a code name, and I may have learned to attach it to a different visual impression from the one you have. We don't know how to find out whether or not that's true, nor do we know which (if either of us) is seeing the colour of the chair-as-it-is-in-itself.

Let's suppose, however, that I say the chair is brown and so is the bookcase, and you agree. Now we haven't settled upon what the chair looks like colour-wise in the ultimate sense, but we have settled upon a relationship, in this case an instance of sameness. Most of us would say that a little objectivity has entered the picture. It's a paradox that at the same time there's a sense in which we have begun to ignore the question of the ultimate objective reality of the colour of the chair. Why worry about the unknowable? We've decided that there is something we can agree on. It's no longer a personal decision. It took two of us. We can predict that if a third person joins us he or she will also say that the chair and the bookcase are the same colour and will probably call it 'brown'. If that happens, so much the better. Science refuses to accept evidence that has no possibility of corroboration, and so the establishment of scientific fact is always going to be a social rather than an individual achievement. But we still don't know the colour of the chair-as-it-is-in-itself, nor is it at all certain that we ever will learn by this process.

Such agreements have carried us far beyond chairs and bookcases, and on other levels as well we've learned not to expect absolutes. For example, no scientific theory we have at present can tell us why the speed of light and the strengths of the fundamental forces of nature are what they are. We've observed the speed of light and the strengths of the forces. Because this is knowledge we can *only* get from observation, we might argue that discovering these values has had nothing to do with any agreement, it is basic knowledge of independent reality. Moreover, we describe them with a number, not a name. A number is a more precise label. 'Two' is 'two' for you and for me with a precision 'brown' can never have.

But what does a number tell us? A number actually reflects the relationship with other numbers, and thus we are able to relate the observed strength of one force to the observed strengths of the others. Scientists would dearly love to be able to explain why these values and relationships should be the ones that apply in our universe – or even show why the ones we observe are more probable than others. There are on-going efforts to understand some of the constants of nature in systems of symmetry. Again, we are concentrating on what we have the capacity to handle – relationships. Absolute values, absolute position in space or time – in the most fundamental sense of the word absolute – science has all but abandoned these notions. Perhaps to God there are such absolute values and positions. We don't even know exactly what that would mean.

How far can the process of discovering relationships take us? To ultimate truth about *relationships* in the universe? Maybe. To ultimate objective knowledge about the universe and beyond? That depends upon whether the relationships are the ultimate reality. We hope that a theory of everything will eventually wrap all of it up – fundamental laws and particles, initial conditions, the constants of nature. But even then, would complete self-consistency, a perfect system of relationships, represent anything absolute, anything beyond 'self-relationship' – a description which scratches its own back to perfection but may not be a description of bed-rock, absolute, chair-in-itself reality?

There are questions that we may never be able to answer with the scientific method, at least as we know it, no matter how successfully we realize its potential. In more than one sense, a scientific Theory of Everything would not necessarily be a Theory of *Everything*.

IS THERE ANYTHING ELSE?

For starters, we must remind ourselves that a scientific Theory of Everything would not allow us to know or predict specifically everything that happens in the physical universe. We are limited by the breakdown of predictability where our observations meet the quantum level. We are limited in our inability to do calculations of great enough complexity and to have available for our calculations the infinite details on any level which would be necessary to predict anything specific. We are limited by a lack of understanding about the relationships between levels of complexity: is everything about molecules determined by what happens on the level of atoms? . . . and so forth. We are limited by a random step in the process of evolution that rules out precise prediction of what creatures will evolve.

Those are severe practical limitations on the value of a scientific Theory of Everything for making specific predictions. Nevertheless, such a theory might *explain* everything by providing a simple formula that would allow for, and underlie, all we observe – without strictly determining the details or allowing us to predict. Let's suppose for the sake of this discussion that some day in the future we do arrive at a complete physical explanation for the universe. Would that be everything there is to know about the universe? Does ultimate truth include anything beyond that ultimate physical and mathematical explanation?

We needn't get spooky about it. Part of the 'anything else' might be human minds and personalities. Can we entirely account for our self-awareness, our minds, personalities, intuitions, and emotions, by means of a physical explanation? This is a matter of enormous significance for many of the questions we are asking in this book, and we will return to it frequently. If we are super-complex computing machines – the sum of our physical parts and their mechanical workings, which in turn exist as a result of the process of evolution – then science may ultimately be able to tell us everything there is to know about us. Even if no computer can ever simulate the human mind, science may find another complete physical explanation. But we have at present no scientific reason to rule out the possibility that there is more to self-awareness, our minds, and our personalities than any such explanation can encompass. Is there such a thing as the soul? If there is, does its existence begin and end with our material existence? Despite some impressive advances in the field of artificial intelligence, and an increasing understanding of the way our minds work, certainly no-one would claim to be able to say at present, except on faith, whether science will eventually be able

ABOVE: The funeral of Charles Darwin, Westminster Abbey, 26 April 1882, with leaders of Science, Government and Church in attendance. The *Manchester Guardian* reassured its readers that they should not have 'any misgivings lest the sacred pavement of the Abbey should cover a secret enemy of the Faith'. (*The Mansell Collection*)

BELOW: A ticket for a better seat at the funeral.
(*Cambridge University Library*)

FUNERAL OF MR. DARWIN.

WESTMINSTER ABBEY,

Wednesday, April 26th, 1882.

AT 12 O'CLOCK PRECISELY.

Admit the Bearer at Eleven o'clock to the

JERUSALEM CHAMBER.

(Entrance by Dean's Yard.)

G. G. BRADLEY, D.D.

Dean.

N.B.—No Person will be admitted except in mourning.

RIGHT: My grandfather and grandmother, Revd A.R. Vetter and Molly Vetter. My grandfather preached in English and German in small Methodist churches in the Texas hill-country. (*By courtesy of the author*)

BELOW: My grandparents' dining-room chair – 'as-it-is-in-itself' perhaps? (*By courtesy of the author*)

LEFT: Jim Morgan. The leaf-hopper did not stay for the photograph, having an appointment elsewhere which had been jotted in his diary at the origin of the universe (see p.30). (*By courtesy of the author*)

BELOW: Stephen Hawking: 'What is it that breathes fire into the equations and makes a universe for them to describe?' (© *Miriam Berkley*)

GEOMETRY AND SYMMETRY
IN NATURE AND IN HUMAN DESIGN:

ABOVE: The seventeenth-century dome of the Chapel of the Holy Shroud, Turin, designed by the architect Guarino Guarini. (*The Mansell Collection*)

OPPOSITE TOP LEFT: Crystals of aragonite.
(© *David Bayliss, RIDA Photo Library*)

OPPOSITE TOP: The entirely natural Giant's Causeway, Co. Antrim.
(© *David Bayliss, RIDA Photo Library*)

OPPOSITE BOTTOM: Alhazen's concept of a cone of rays coming to the eye from an object is the foundation of perspective. Albrecht Dürer woodcut 'Demonstration of Perspective' from *Elementa Geometrica*, 1525.

ABOVE: Seeking the Holy Grail, Monty Python's bumbling knights ride imaginary horses.
(© *Python/EMI. Courtesy of the Kobal Collection*)

RIGHT: Designer Harry Kupfer's laser-lit gods, with gleaming plexiglass weapons and lucite suitcases full of invisible power, celebrate their arrival at Valhalla in Richard Wagner's *Das Rheingold* at Bayreuth.
(© *Bayreuther Festspiele GmbH/Rauh*)

DISCOVERING THE *W* AND *Z* PARTICLES OF THE ELECTROWEAK FORCE:

ABOVE: The UA1 detector at CERN. (© *CERN, Geneva*)

RIGHT: The first recorded instance of the production of the Z particle, 1984, found in the debris from a proton-antiproton collision. The two white arrows point to the tracks of an electron and positron emerging in a manner which indicates that a Z particle has decayed. The W^+, W^-, and the Z are too short-lived to leave detectable tracks of their own. (© *CERN, Geneva*)

ABOVE: Vesto Slipfer at the Lowell Observatory, *c*.1912. (*Lowell Observatory photograph*)

ABOVE RIGHT: Edwin Hubble in the 1920s, with the 100" telescope at the Mount Wilson Observatory. (*Henry E. Huntington Library and Art Gallery*)

RIGHT: Robert Wilson (on ladder) and Arno Penzias inside the horn antenna at Bell Laboratories in New Jersey, 1965. (*AT&T Archives*)

to assimilate the phenomena of self-awareness, mind, and personality into the materialistic picture. If science can't, then there is truth beyond the range of scientific explanation.

Another part of the 'anything else' may be what we call the supernatural. Perhaps it is simply figments of imagination, psychological events, not so much to be explained by science as to be explained away. Or perhaps these are real events which are at present unexplainable because we lack complete understanding of the full potential of the physical world. If either is the case, then the supernatural ought eventually to fall into the realm of scientific explanation. However, if the supernatural world exists, and if it is inherently beyond testing by the scientific method, then there is truth beyond the range of scientific explanation. There may indeed be more in heaven and earth than is dreamed of in our science (if not our philosophy).

Another part of the 'anything else' may be the 'meaning' of what we experience. There's meaning in the sense of the significance you or I attach to a physical event, and there's meaning in the sense of ultimate significance which doesn't depend upon our recognizing it. If our psychology is entirely explainable in terms of physical processes, as we said above it might be, then any meaning you or I attach to events might be similarly explained by science. For instance, the birth of my child may have meaning for me beyond the physical event because of my psychology and chemistry. Suffering, beauty, evil might all be reduced to physics, chemistry, and the way we have evolved to feel, think, and react. Perhaps there is no meaning in any of this beyond the ability of science to explain.

But there is no *scientific* reason to rule out the possibility that a birth has meaning in an ultimate sense. Is life sacred? If so, why? Because the child is known to God? The possible sacredness of life is a 'meaning' that the scientific method can't explore. Is there evil which is evil or beauty which is beauty in an ultimate sense, not subject to human interpretation or traceable merely to the way we have evolved to prefer things? Does any event have a meaning in the sense that Christians claim when they say that a crucifixion brought the possibility of universal salvation? If meaning is more than human-made significance, interpretation, and symbolism, then there is truth beyond the range of scientific explanation.

Another part of the 'anything else' may be God or some other answer to the 'Why' of the universe. If the Mind of God is only a euphemism for the sum of all the laws of physics, then God is not beyond the reach of science. But Hawking has written: 'Even if there is only one possible unified theory, it is just a set of rules and equations. What is it that breathes fire into the equations and makes a universe for them to describe? The usual approach of science of constructing a mathematical model cannot answer the question of why there should be a universe for the model to describe.'[44] Some of his colleagues would disagree with him, but if Hawking is right, then there is truth beyond the range of scientific explanation.

But aren't we being a little far-out? There is a philosophy which has it that if science cannot study something, that something cannot be real. That may seem extreme, but it receives some support from the supposition that the limitations of our measuring capacity on the quantum level are actually limitations on what can take place there. On the face of it, it may seem ridiculous to carry this supposition into other areas of science and human experience, but the idea crops up regularly in both science and religion.

However, hasn't science proved to us in more positive ways than 'Sorry, can't study it' that the supernatural world is only a trick of the brain, only psychological experiences, at most unusual but altogether natural occurrences? Hasn't it shown that what we call God is only the laws of physics, or wishful thinking? Hasn't it shown that meaning is only interpretation – meaning in the eye of the beholder? And isn't there already good evidence that human mind and personality are only the product of complex physical mechanisms?

No. Science has not yet been able to offer us a complete physical explanation in any of these four areas; we do not know that it has the capacity ever to do so; we do not know whether there are, even in principle, *unknowable* physical explanations. But even were science to give us a complete physical explanation, we couldn't claim to have found the ONLY explanation. We couldn't even claim to have found the only complete *physical* explanation. In fact, we couldn't even claim to know we have found the simplest or the best physical explanation. Present-day science knows no way of showing that we have, only that we haven't – by finding a simpler or better one. To say that something has been 'proved' to be 'only' this, or 'only' that, is not a scientific statement. However, although science can't prove there are no alternative or better explanations, we're going to see that it *can* show that some of the alternative explanations are unsatisfactory and inconsistent with the evidence. It is not a situation of anything goes.

Meanwhile, are we hedging our bets here to the point of the ridiculous, clutching at straws and making things more complicated than necessary to keep open the possibility of God, meaning, the supernatural, and the human soul? If we have explained something satisfactorily, perhaps we *could* look for another explanation; but why *should* we?

If there is any hope at all of perceiving the world without being entangled by a point of view, the realization of that hope must surely begin with a lesson we could have learned from everyday experience or almost any good fictional detective story: A simple explanation that ties in all the evidence isn't necessarily the right explanation. The black creature in the tree is not always a crow. The loud bang heard one summer day in my neighbourhood was not a backfire or a firecracker, or even someone murdering his wife. It was my neighbour shooting golf balls at a tree from a home-made brass mini-cannon. Real life, especially but not exclusively where human behaviour is involved, is not always best represented by the simplest explanation.

We believe that there are uncomplicated, elegant laws underlying the untidiness. But even in science there always seem to be more loose ends than most of us prefer to think, and there is what fictional detective Sir Henry Merrivale[45] called 'the blasted cussedness of things in general' – the propensity of things to lend themselves to a simple, meaningful explanation, when in fact they have come about in damnably complicated and illogical ways. Attempts to generalize, to explain in a manner that satisfies our desire for neatness, simplicity, the most logical chain of cause and effect, may not always lead to truth. We can't be sure that ultimate truth *will* be simple. And of course there's another possibility, that our simplest scientific explanation isn't simple enough – that Bernstein was right about God being the simplest of all.

Have we in these paragraphs demolished the argument that

everything is explainable by science in physical terms? Certainly we have done nothing of the sort. We've shown that science can't prove that a physical explanation is THE complete explanation. We haven't in turn shown that there *is* another explanation, or anything else to explain.

THE INSIDIOUSNESS OF GOD

The old, pre-Darwin 'natural theology' was a search for evidence of God in the works of his creation. Because science has found other explanations for the origin of so much that used to be considered explainable only as the work of God, there seems little basis for faith left in natural theology. We can no longer declare that nature points irresistibly beyond itself. However, the philosophical questions raised by science *do* irresistibly point beyond science. It is not without reason that Hawking said: 'It is difficult to discuss the beginning of the universe without mentioning the concept of God.'[46] Fred Hoyle wrote: 'I have always thought it curious that, while most scientists claim to eschew religion, it actually dominates their thoughts more than it does the clergy.'[47] Perhaps when C. S. Lewis warned that 'a young man who wishes to remain a sound atheist can't be too careful of his reading',[48] the reading to be strictly avoided should include science books.

A common reaction of scientists making a new discovery about how the universe works is 'How clever! I would never have thought to do it that way myself!' The next thought that springs to mind is 'Who did?' Is it wisdom or naivete or social conditioning that brings that thought? Is it a God-given instinct that makes us ask 'Who?' when there is a flash of recognition that here is a mind like our own but far superior? Or are we so hopelessly anthropomorphic that we have difficulty allowing a clever pattern to be the first cause of everything rather than a clever person?

Our spontaneous 'Who did?' is not the only question which points beyond science.

THE MORALITY OF SCIENCE: IS TRUTH GOOD?

The visceral feeling among most scientists and many of the rest of us is a little at odds with the principle that scientific findings themselves have no moral content, that 'truth' does not imply 'good'. There is a moral arrow in science aside from the question of how findings are put to use or whether we approve of them. It is a feeling first that it is good to do science – more worthwhile than many other pursuits a person might undertake. We consider uncovering truth about the universe a high calling. Beyond that is a feeling that truth is inherently better than falsehood, knowledge is better than ignorance, objective truth will be beautiful and orderly, not ugly and confused. Whether or not there was a God at the Creation who 'saw that it was good', we assume that objective truth has a purity, a healthy feel about it. There are only a few who think that ultimate truth might be horrible, utter confusion, madness. Instead, following the arrow toward objective reality seems to remove us from the nitty-gritty fallen world where men and women use this raw material of reality so unwisely – removes us to holier ground, nearer the ultimate 'good', the Mind of God or the perfection of human knowledge.

Little arrow, who made thee? Or what? The moral arrow in science which defines a direction toward truth and away from falsehood, toward knowledge and away from ignorance, toward beauty and away from ugliness, attaching a value to these directions – this 'arrow' is not easy to explain, though many have tried. It seems to come from instinct. Perhaps it's the result of evolution. Does this sort of thinking give a survival advantage? Is 'good' only what was 'good' for the species? Have our minds over-evolved to attach an aesthetic and even moral value to the compression of information into simplified patterns? Perhaps the arrow results from our cultural conditioning or is a 'meaning' of our own invention because it pleases us to think of things this way. Is it part of the uncaused laws of the universe – as Hardy says '317 is a prime number' is? Is it perhaps nothing more than a rationalization for doing science which makes science more than just another way to eke out a living? Is it a pointer to anything at all?

Some say that the question about why the arrow exists, about who or what has set this compass, can best be answered by saying there is a God whose nature defines 'truth' and 'good' and 'beauty'. C. S. Lewis applied this argument not just to science but to the question of why there is any good/evil orientation in the universe at all.

Nature fills us with delight and awe. It moves us profoundly in ways that are difficult to express or assess and leads us to ask questions science may never be able to answer. But does it point to God? Before Darwin, many of our forebears had no philosophical misgivings about singing a hymn set to the music of Haydn whose last stanza, after admitting that the stars and planets have no voices in the usual sense, nevertheless says that 'In reason's ear', they unmistakably declare, 'the hand that made us is divine.'[49] What sort of reason would we have to employ to hear that declaration today?

4

ROMANCING THE CREATION

The evolution of the world may be compared to a display of fireworks that has just ended: some few red wisps, ashes and smoke. Standing on a cooled cinder, we see the slow fading of the suns, and we try to recall the vanished brilliance of the origin of the worlds.

GEORGES LEMAÎTRE

OUR LATE TWENTIETH-CENTURY PICTURE OF THE UNIVERSE IS dramatically different from the picture our forebears had at the beginning of the century. Today it's common knowledge that all the individual stars we see with the naked eye are only the stars of our home galaxy, the Milky Way, and that the Milky Way is only one among many billions of galaxies. It's also common knowledge that the universe isn't eternal but had a beginning ten to twenty billion years ago, and that it is expanding. We take all this so much for granted now that it's hard to believe how far we've come in the past ninety years in the quest to discover the origin of the universe.

In spite of our greater understanding, the universe has become in many ways even more mysterious to us than it was to earlier generations. It is not a familiar, cosy place. It stretches out to distances inconceivably vast and contains systems driven by incredible power. Earth now seems tiny and insignificant, a speck, a cooled cinder. It would appear that if we humans are of any interest to the Mind of God, God carries to an absurd extreme the credo of Dr Seuss's elephant Horton: 'A person's a person, no matter how small.'[1]

The first part of this chapter is a short review of the chain of theoretical and observational discovery which led over a period of years to the conclusion that the universe began with a Big Bang. We will also look at the philosophical and religious controversy which greeted these astounding and sometimes unwelcome developments. Those to whom this story is already familiar may want to move quickly through these pages to the middle of the chapter and more contemporary debates.

THE UNCOMFORTABLE CONCEPT OF A BEGINNING

By the end of the First World War there was no concrete evidence that the turn-of-the-century picture of the universe was incorrect, but there were suspicions. Since the eighteenth century there had been speculation about fuzzy patches of brightness called the nebulae. It seemed most likely they were only gas clouds in our galaxy, but some people entertained wilder ideas: they might be newborn solar systems, or fissures in the universe where matter and energy pour in from another universe or another dimension, or remote, independent formations of stars and gases like the Milky Way. Perhaps the Milky Way was only one among many 'island universes'.

In the early years of the twentieth century, attention had begun to focus on those nebulae that had a spiral structure, because many astronomers thought these were protostars – clouds of collapsing gas on the point of giving birth to a star. Between 1912 and 1914, Vesto Slipher at the Lowell Observatory in Flagstaff, Arizona, discovered that most of the spiral nebulae he was studying showed a red shift: that is, a shift in the colours of the spectrum of light away from the blue end of the spectrum and toward the red end. Slipher interpreted this shift in the light coming from the nebulae to mean that the distance between us and them was growing greater, just as we interpret the drop in the pitch of an engine or siren to mean that a vehicle is moving away from us – the familiar Doppler effect. In both cases the shift is caused by the stretching of waves that reach us from something as its distance from us increases. In the case of the siren, sound waves are stretched. Our ears interpret the length of a sound wave as pitch; we 'hear' longer sound waves as a lower pitch. In the case of the spiral nebulae, light waves are stretched. Our eyes interpret the different lengths of light waves as different colours, and longer light waves mean a shift to the red end of the spectrum. The sort of red shift Slipher was discovering is not detectable to the naked eye as reddening light. He based his conclusions on calculations made by studying the spectra of light from the nebulae and comparing them with the spectrum of light from something whose distance from us is not changing.

What Slipher had found was revolutionary. In 1914 he presented his findings to the American Astronomical Society. John Miller, who had been one of Slipher's professors, described the event: 'Something happened which I have never seen before or since at a scientific meeting. Everyone stood up and cheered.[2] The turn-of-the-century picture of the universe was on the brink of crumbling.

Clearly Slipher had made a discovery of enormous importance, but it wasn't immediately obvious what it meant. Slipher's interpretation was that our own drift through space was causing the increasing distance between us and the nebulae. Since we don't think of the universe in terms of absolute position, it might seem a moot point who is retreating from whom, but Slipher's 'drift' didn't take into account the more dramatic implications of his discovery. Those wouldn't emerge until many more observations had been catalogued.

One problem with interpreting the significance of the red shift was that no-one was yet able to determine how far away the spiral nebulae were. The difficulty with measuring distances to objects in space is similar to the difficulty we have in judging the distance between ourselves and a light shining at night: is the light a few feet away and very faint, or is it a few miles away and very bright? Though the distance of the nebulae was still in question at the time of Slipher's announcement, astronomers were not far from having the answer. Since the last decade of the nineteenth century, they'd been devising increasingly sophisticated ways of measuring such distances.

Meanwhile, what were the theorists saying? Einstein produced his Theory of General Relativity in 1915. Within the next two years, Dutch astronomer Willem de Sitter and Einstein himself began to see that solutions to Einstein's equations implied that the universe is expanding. Einstein, like most of his contemporaries, believed the universe is static, that is, not changing in size. When the implications of his equations began to emerge, he was chagrined. As he wrote in a letter, 'To admit such a possibility seems senseless.'[3] He decided to adjust his theory to cancel out the prediction of an expanding universe by putting in a new constant of nature – a 'cosmological constant', a mathematical term which corresponded to a force of repulsion or 'anti-gravity'. Einstein was later to dub this cosmological term – this concession to his own preconception and that of his contemporaries – 'the biggest blunder of my life'.

The Russian mathematician Alexander Friedmann was the first to buck the spirit of the times emphatically and insist on taking Einstein's theory at face value, not assuming that the 'cosmological term', if it had to be considered at all, was necessarily anything other than zero. What Friedmann found was not just one solution but a family of solutions to the cosmological equations of General Relativity, and each different solution describes a different sort of universe.

The Belgian astrophysicist and theologian Abbé Georges Henri Lemaître – with whose words we opened this chapter – found solutions to Einstein's equations which were similar to Friedmann's. However, unlike Friedmann, Lemaître was most intrigued with what the equations and their solutions could tell him about the origin of the universe. It was he who first envisioned something like what we now call the Big Bang, though he didn't give it that name. Partly because he was a priest as well as an astrophysicist, this idea was met with some derision from fellow scientists. Lemaître's suggestion was that there had been a time when everything that makes up the present universe was compressed into a space only about thirty times the size of our sun – a 'primeval atom'. As he put it, 'The primeval atom hypothesis . . . pictures the present universe as the result of the radioactive disintegration of an atom.'[4] By the time Lemaître wrote those words in the fifties, he was speculating that this primeval atom might be thought of as a single quantum.

While Friedmann's theoretical work remained largely

unknown except among mathematicians – he died in obscurity at the age of thirty-seven – Lemaître's gained the attention of observational astronomers, largely thanks to Eddington (whose student Lemaître had been at Cambridge) and another of Eddington's research students, George McVittie.

Meanwhile, back in Arizona, Vesto Slipher continued to design his own instruments for studying the nebulae and discovered that most he was able to study showed red shifts. In early 1921 he reported an enormous red shift (or what seemed enormous at the time) for a nebula called NGC584. According to Slipher's calculations the nebula's distance was increasing at a speed of approximately two thousand kilometres per second. In 1922 Slipher sent Eddington at Cambridge measurements for forty spiral nebulae, thirty-six of which were receding.

When Slipher first announced his findings about red shifts in 1914, a young man named Edwin Hubble had been in the audience. In the years that followed, Hubble began to see the connection between Slipher's observational discoveries and the solutions that de Sitter (and Lemaître and Friedmann – though Hubble may not then have known about their work) was getting from Einstein's equations. Hubble also turned his attention to the nebulae. In 1923 he realized that a faint spot of light in the Great Nebula in Andromeda was not a nova, as he had previously thought, but a Cepheid – a star that regularly changes its brightness. It was this realization that enabled him finally to settle the question whether the nebulae are something in our galaxy or remote, independent 'island universes'. Astronomers had learned how to calculate the distance to a Cepheid by timing these variations. Hubble's calculations showed that the Andromeda nebula is at a distance much greater than any star in the Milky Way. It is indeed another galaxy.

Hubble went on to establish that there are many galaxies besides our own, and in 1929 he made one of the most revolutionary announcements in the history of science, one that was to change forever our ideas about what the universe is like, about its history, and about ourselves. He and his associate Milton Humason, a colourful character who had begun not as a scientist but as a mule driver at the Mount Wilson Observatory, established that except for galaxies that are clustered closest to us every galaxy in the universe is increasing in distance from us.

Moreover, except for galaxies which are close together, every galaxy in the universe is increasing in distance from every other galaxy.

The observations continued, and more and more galaxies and red shifts were catalogued. By the early fifties the relationship between what astronomers were discovering with their telescopes and the theoretical predictions of Einstein, Friedmann, and Lemaître was clear. The red shifts become greater the farther away a galaxy is from us, which tells us that the farther away the galaxy is, the faster it's receding. As Friedmann had predicted, regardless of where we were to station ourselves in the universe, in any galaxy, we would see the other galaxies receding from us, twice as far away, twice as fast. A loaf of raisin bread rising in the oven is a homely analogy to illustrate this. Standing on any raisin while the dough rises and expands between the raisins, we would see every other raisin moving away from us – twice as far away, twice as fast. The raisin bread also reminds us that it is more accurate to think of the expansion of the universe, as Friedmann first suggested, not in terms of galaxies flying away from one another through space, but in terms of the space between them swelling.

One might easily jump to the conclusion that if the universe is expanding like a loaf of raisin bread, we ought, if we had the technology to do so, to be able to travel to the surface of the loaf and find the edge of the universe. What would be beyond? The question of what is beyond the edge unfortunately has no real meaning. Eddington suggested that we think of a balloon with dots painted on its surface. Imagine an ant crawling on the surface of the balloon. In order for the analogy to be helpful we must say that for this ant all that exists is the surface of the balloon. The ant can't look outward from the balloon's surface or conceive of an interior to the balloon. Those dimensions don't exist for the ant. Now if air is let into the balloon and the balloon expands, the ant will see every dot on the surface of the balloon moving away from it. Regardless of where the ant travels on the balloon, every dot will be moving away. The ant won't find an edge or an end anywhere. The same may be true in our universe, but with more dimensions than in the ant's balloon universe.

Another conclusion to which we might jump is that we ought to be asking where in the universe the expansion began. Where is

94

the point everything is retreating from? One way of thinking of the expansion of the universe is as an explosion outward. Even if there are no absolute directions in the universe, beings riding on a piece of debris from any explosion ought to be able to assume that there is an answer to the question: Where exactly did the explosion take place in relation to where we are now? Eddington's balloon analogy helps us understand why there is no such point of origin in the universe. On the balloon surface, there is no such point – or, if you prefer, any point could just as fairly claim to be the point of origin. Remember that the interior of the balloon is a dimension that doesn't exist. Modern cosmology accepts Friedmann's assumptions: the universe looks the same (on the large scale) in all directions; and regardless of where we were to stand in the universe it would look the same in all directions. There is no edge from which we would see galaxies in one direction and nothing in the other. There is no core toward which we could point and say, There it began.

We can, however, ask *when* the universe began.

Any direction in space we look, no matter where in the universe we are, we look toward the past. Even in so small a space as the room where I sit and write, what I see is old news. However, the delay with which the picture of the far wall reaches my eyes is not worth considering, because light – and thus any picture that comes into my eyes – does travel extremely fast.

When it comes to cosmic distances, the delay is decidedly worth considering. The light that reaches us from some distant quasars left them perhaps ten billion years ago.[5] Are the quasars still there? In give-or-take another ten billion years our descendants on the earth (if descendants and earth still exist) might find out whether these quasars, or the galaxies into which they may have evolved, were still there in the 1990s (earth time). From our own vantage point, we can only observe their existence ten billion years ago. Since the past is in all directions, then out there – some distance beyond the quasars – is the answer to the questions: Did the universe have a beginning, and, if so, when?

Fortunately, there are other ways of finding the answers to those questions besides actually seeing the split second of the origin of the universe – an observation which is not possible with our technology and perhaps not with any we could ever invent. If the universe is expanding, it would seem correct to think that it must at an earlier time have been, as Lemaître insisted, much denser than it is now. In fact it would seem correct to think that there was a time when everything we would ever be able to observe in the universe was in exactly the same place, and that this must have been the beginning.

Must that have been the case?

In 1948 Hermann Bondi, Thomas Gold, and Fred Hoyle introduced theories which allowed for the expansion of the universe but did away with the requirement that the universe must have a beginning. According to their 'Steady State' theory, the universe hasn't always contained all the matter it does today. As the universe expands, new matter continuously emerges to fill in the gaps, and the average density of matter in the universe remains the same. Galaxies such as ours reach the end of their life cycles – when the stars in them burn out and the galaxies die – but meanwhile new galaxies are forming from new matter.

A Steady State universe would have no beginning or end. This return to the possibility of an eternal universe was welcomed by many, including the theorists who invented it, as a way of eliminating the hint of 'creation' that was inherent in a universe with a beginning. For more than a decade the scientific and (to a lesser extent) the philosophical debate continued between those who favoured the Steady State theory and those who favoured the Big Bang.

It may be difficult from our vantage point to understand why the notion of a beginning presented a major philosophical problem for anyone. Today almost all scientists accept some version of the Big Bang theory, yet we still find atheists and agnostics as well as believers in God among them. Clearly having a Big Bang must not prove decisively that we have a God. As we will see a little later, having a Big Bang doesn't even prove we have a beginning. Why were Bondi, Gold, Hoyle, and some of their colleagues so concerned? We must try to see this from the point of view of those who debated it in the late forties and the fifties.

To a certain extent it was true that as the Big Bang theory began to look increasingly likely to be the correct one, the anti-God camp seemed to be losing ground to the pro-God camp, but that was not the whole story. In Chapter 3 we saw how Robert Jastrow, himself an astronomer and an agnostic, in his book *God and the Astronomers*, chides his fellow scientists for their reaction to the Big Bang theory: 'the response of the scientific mind – supposedly a very objective mind – when evidence uncovered by science itself leads to a conflict with the articles of faith in our profession.' Jastrow describes the situation:

> This is an exceedingly strange development, unexpected by all but the theologians. They have always accepted the word of the Bible: In the beginning God created heaven and earth. To which St Augustine added, 'Who can understand this mystery or explain it to others?' The development is unexpected because science has had such extraordinary success in tracing the chain of cause and effect backward in time . . .
>
> Now we would like to pursue that inquiry farther back in time, but the barrier to further progress seems insurmountable. It is not a matter of another year, another decade of work, another measurement, or another theory; at this moment it seems as though science will never be able to raise the curtain on the mystery of creation. For the scientist who has lived by his faith in the power of reason, the story ends like a bad dream. He has scaled the mountains of ignorance, he is about to conquer the highest peak; as he pulls himself over the final rock, he is greeted by a band of theologians who have been sitting there for centuries.[6]

However, as Jastrow himself pointed out, the controversy was much more complicated than a simple competition between science and religion in which religion had apparently won a major victory. It isn't God that Jastrow's scientists find when they pull themselves over the final rock. It is a band of people, including presumably St Augustine, faced with a closed door at a beginning in time through which we are not allowed to pass in our search to know everything.

The irony in Jastrow's story is not that the theologians have had it all explained for a long time, while the scientists have not. The irony is rather that the theologians have been saying for many centuries that we are dealing with a mystery human beings will *never* be able to explain, and now the scientists, by dint of hard labour trying to find that explanation, have to their chagrin arrived at the same conclusion. It isn't the discovery of God, but rather the discovery of the limits of human intellectual endeavour that rots everyone's socks on the mountaintop. The theologians have learned to live fairly comfortably with those limitations and put down roots and even enjoy the situation. The advantage they claim to have, and if it's true it is a very great advantage, is that they believe the end of human intellectual endeavour isn't necessarily the end of the quest for complete understanding.

For a while the Steady State theory that allowed one to believe that the universe was eternal held its own and seemed a powerful rival to the Big Bang theory. Both theories seemed equally capable of explaining what had been found by observation. However, in the sixties, new evidence came to light which the Steady State theory could not explain and the Big Bang theory could.

Back in the 1940s, George Gamow, a Russian-born physicist who defected to the West in 1933, had begun, with Americans Ralph Alpher and Robert Herman, to theorize about the early universe by running Friedmann's equations backward toward the event with which the universe began. They predicted that there should be left-over radiation – photons (messenger particles of the electromagnetic force) – surviving from about a thousand years after the origin of the universe. In that era the universe would still have been very hot, but the prediction was that the temperature of those photons should by now have cooled to about five degrees above absolute zero. Such radiation would be very difficult to observe, and the prediction was not tested. The evidence of that radiation was finally discovered by accident in 1965. The story of the discovery recalls our discussion of the interplay between theory and direct observation in Chapter 3. It is an instance in which theory didn't lead the way but rushed in with the spectacles needed to make sense out of otherwise puzzling data.

In the mid-1960s, at Bell Laboratories in New Jersey, there was a horn antenna designed to be used with the Echo I and Telstar communication satellites. The amount of background noise the antenna picked up hampered its use in the study of signals from space. Scientists working with the antenna had to make adjustments and confine themselves to studying signals that were stronger than the noise. It was an annoyance that was possible to ignore, but two young scientists, Arno Penzias and Robert

Wilson, took the noise more seriously. They noticed that the noise remained the same no matter which direction they pointed the antenna. If the noise were a result of the earth's atmosphere, that wouldn't be the case, since an antenna pointed toward the horizon faces more of the earth's atmosphere than one pointed straight up. The noise had to be coming either from beyond the earth's atmosphere or from the antenna itself. Wilson and Penzias thought pigeons nesting in the antenna might be causing the disturbance, but evicting the pigeons and clearing away their droppings made no difference in the noise.

Wilson and Penzias weren't aware of a current proposal from Robert Dicke at Princeton, who was in the process of building an antenna to search for the background radiation that Gamow, Alpher, and Herman had predicted in the 1940s. But when another radio astronomer, Bernard Burke, heard from Penzias and Wilson about their problem with the antenna, he proceeded to bring the two groups of researchers together. Penzias and Wilson had found by accident the radiation that Dicke, led by theory, had been hoping to find.

In 1973, balloon experiments of Paul Richards and others at Berkeley in California showed that the spectrum of the background radiation was the spectrum Big Bang theory predicted. The cosmic background radiation (as it is now called) has been confirmed by many experiments and is the most direct evidence we have that the universe was once very much hotter and denser than it is now. The radiation as it reaches us has a temperature of about three degrees above absolute zero, instead of the five degrees Alpher and Herman had calculated. Today we know that you don't need unusual equipment to observe the cosmic background radiation. The snow on a TV screen that appears when a station isn't broadcasting consists in part of this radiation – these photons which are artefacts of ancient light.

The discovery of the cosmic background radiation and its spectrum was dramatic support for the Big Bang theory. There was more. The theory predicts that, of all the elements making up the universe, about 25 per cent of the mass ought to be helium 4. By the mid-seventies, measurements of the elements in external galaxies (a measurement which is possible by studying their spectra), as well as in our own galaxy, confirmed this prediction. They also confirm predictions of abundances of other elements that were made in the Big Bang, such as deuterium, helium 3, and lithium.

More support for the theory came from the fact that it suggests a solution to the mystery of why we find quasars only at such large distances from us. Most astrophysicists link quasars with galaxy formation. If galaxies were periodically dying and being replaced by new galaxies made from new matter, as the Steady State theory would have it, then we ought to find quasars fairly evenly scattered throughout the universe. On the contrary, we find no quasars near us. They are all far away, and, by virtue of that fact, long ago. It's understandable why this is so if galaxy formation occurred mainly during one era far back in the history of the universe, and is not a continually recurring process. Looking to the distance where the quasars are, we are seeing the universe in that era of galaxy formation. The information from there has taken a long time to reach us. Old news indeed, but it seems to indicate that we are in a universe that is evolving over time, a universe like the Big Bang universe, not the Steady State universe.

While observational evidence was confirming the Big Bang, theorists were providing further support and putting an additional padlock on the slammed door at the beginning of time. It had become clear that if general relativity is correct, it's overwhelmingly probable that the universe will be either expanding or contracting. A static universe in that theory is about as stable as a pencil standing on end. Nevertheless the question arose, If a universe is expanding, even if it isn't a Steady State universe, does that necessarily mean that everything in it was in the same place at some earlier time?

In 1963 Russian scientists Evgenii Lifshitz and Isaac Khalatnikov suggested another possible history for an expanding universe. Running time backward, imagine a scenario in which a universe something like ours contracts, with all its galaxies getting closer together, apparently on collision course. Looking more closely at the galaxies, we notice that they have other motion in addition to the motion that's drawing them directly toward one another. When the galaxies approach one another, this additional motion might cause them to miss one another, fly past – and the universe expand again without having reached a state of infinite density.

It was this possibility that interested Hawking and Roger

Penrose in the middle and late 1960s, about the same time Wilson and Penzias were puzzling over the cosmic background radiation. General relativity predicts the existence of singularities – points of infinite density and infinite spacetime curvature – but in the early sixties few physicists took this prediction seriously. Some thought that a star of great enough mass undergoing gravitational collapse might form a singularity at the centre of a black hole. No-one yet had claimed that it MUST.

Though some of John Wheeler's students say they heard him use the words earlier, 1967 is usually the date given for his coining the term 'black hole'. However, the study of black holes began well before that, as we learned in the story about Chandrasekhar in Chapter 3. In 1965 Penrose, building on earlier work of Wheeler, Chandrasekhar, and others, was able to show that if the universe obeys general relativity and several other constraints, when a very massive star has no nuclear fuel left to burn and collapses under the force of its own gravity it will inevitably be crushed to a point of infinite density and infinite spacetime curvature – a singularity. This will happen even if the collapse isn't perfectly smooth and symmetrical. No 'might' about it. It must.

Hawking took off from there. In his doctoral thesis at Cambridge in 1965, he reversed the direction of time and applied the concept to the entire universe. He suspected that what we would see if we could watch the expansion of the universe run backward was similar to what Penrose had found with black holes. Once the collapse (the expansion of the universe run backward) had gone far enough, the additional motions of galaxies would make no difference to the history of the universe. By 1970 Hawking and Penrose were able to show, in Hawking's words, 'that if general relativity is correct, any reasonable model of the universe must start with a singularity'[7], with everything we would ever be able to observe in the universe compressed not to the sphere Lemaître envisioned, but to infinite density. Space-time curvature at the singularity would also be infinite. The distance between all objects in the universe (though calling them objects at this point would be inaccurate) would be zero.

THE GORDIAN KNOT OF SINGULARITY

The slammed door was now locked indeed. Physical theories can't work with infinite numbers. When the theory of general relativity predicts a singularity of infinite density and infinite spacetime curvature, it is also predicting its own breakdown. In fact all the theories of classical physics become useless at a singularity. There's no way to predict what will emerge. Standing at a singularity we can only wait to observe what's to come. In addition, we have no way of finding out why a singularity suddenly ceases to be a singularity and becomes a universe. Any leap of imagination is as good as any other. And what if we turn around to study the past? What happened before the singularity? It's not even clear that these questions have any meaning. A singularity at the beginning of the universe means that the beginning is beyond the limits of our science. All we can say is that time began, because we observe that it did. Hawking and Penrose had tied a true Gordian knot.

The Big Bang scenario for the origin of the universe had come to this:

In the beginning was the Singularity. Everything that was to be the matter/energy of the universe that we might eventually be able to observe was packed together in a point of infinite density. Ten to twenty billion years ago (as 'time' is measured in the space-time frame which was to follow), this 'exploded'. That was the Big Bang. To imagine the infinite heat of 'time zero' of creation is as impossible as imagining the point of infinite density. To imagine the light of it is also a meaningless endeavour, because light as we are able to see it didn't exist. After a time, matter, instead of radiation, began to dominate the universe. The universe expanded and cooled enough for electrons and nuclei to form stable atoms. Matter could begin coming together by dint of its own gravity, starting the process that would eventually lead to stars and galaxies and planets. Ten to twenty billion years after the beginning, we find the universe we know today.

I find myself picturing this process as though I were standing on the outside, watching it take place. But such a position doesn't exist. There was no 'outside' where I could have stood at the beginning, just as there seems to be none today – no vantage point beyond the universe from which to observe the universe.

Everything was within the point of infinite density. Everything was within the explosion. Everything still is.

This was the Big Bang creation story as it existed in the mid-1970s, and on the face of it it was a congenial one for those who believed in God or simply found eternity monotonous and weren't too terribly concerned if humans couldn't know absolutely everything. Both sides of the God-or-not debate – when it has seemed in their interest – have argued with great ingenuity that whether or not there was a Big Bang singularity isn't really relevant to the question of whether or not there is a God. But hardly anyone felt there was nothing at stake in the answer. A very young friend of mine summed it up in a truism: 'If there was a beginning, and we can know about it but we can't ever explain it, that's just a whole different kind of universe.' If you don't like this whole different kind of universe, then the next step is to get busy trying to explain the beginning – or explaining it away.

We have two tracks we must take now to follow this adventure up to the present. In the years since the mid-1970s, theorists and researchers have continued trying to solve problems that still existed with the Big Bang theory. Theorists have also got busy undermining the singularity.

We've said that by the early seventies it was clear that the Big Bang theory could explain much of what we were finding by means of observation, much that the Steady State theory couldn't explain. However, the Big Bang theory could not at that time (nor can it now) explain all the observational evidence. Two of the remaining puzzles have to do with the nature of matter.

First, how can we explain the fact that the universe has matter in it at all rather than being empty? The production of matter is no longer a complete mystery to us. We know how to produce a particle of matter out of pure energy in the laboratory. But we don't know how to do that without at the same time producing an equal amount of antimatter. According to Big Bang theory, a great deal of matter was produced out of energy in the early universe. This raises the question: What has happened to all the antimatter that must have been produced at the same time?

If equal amounts of matter and antimatter appeared in the early universe, as they do in the laboratory, we have every reason to expect that by now there would be neither matter nor antimatter left around, because when matter meets antimatter they annihilate in a burst of pure energy. Every particle of matter would long ago have met an equivalent particle of antimatter and they would all have annihilated each other. The whole game would have ended disappointingly, cancelling out like a card game of Old Maid where the Old Maid card is missing from the pack.

One suggested solution to this puzzle is that most of the antimatter is elsewhere in the universe, while our neighbourhood is an area containing mostly matter. The trouble with this idea is that there would be borders between the regions that had matter in them and the regions that had antimatter in them. It would be difficult not to notice where these borders lay, because matter and antimatter would be annihilating each other there in a way we are able to detect with gamma-ray detectors. So far no such activity has been detected in the region of space accessible to such detectors.

Another suggested explanation goes like this. When matter and antimatter first evolved in the early universe, there was a lot more of it than we see around today, with an imbalance (perhaps very small in proportion to the total amount of matter and anti-matter) in favour of matter particles. After the big annihilation scene, there were left-over matter particles which hadn't found an antimatter partner with which to annihilate. These left-over matter particles, these Old Maid cards, make up all the matter of our universe today. We said in Chapter 2 that for the universe to exist as we know it a certain amount of asymmetry is required. If everything balanced out perfectly and came out even, we wouldn't have the universe. In this explanation for the origin of matter we see a good example of that necessary asymmetry.

If that's the way it happened, we still haven't solved our problem completely. How do we explain the initial imbalance, be it ever so small, between matter and antimatter? Some of the theories which propose to unify the forces of nature provide conditions under which such a situation of imbalance could occur, but so far we have no clear evidence to show which if any of these theories is correct or that these conditions existed. Some of the ingredients are there, but not by any means all. This mystery still remains unsolved.

A second problem concerning matter was, until the spring of 1992, even more of a challenge to Big Bang theory. In repeated measurements researchers had found that the cosmic background radiation is remarkably uniform in temperature. Taking readings out to the end of observability in every direction, they found the temperature the same. This was clear evidence that the early universe was smooth, without lumps, clumps, and irregularities that would show up as fluctuations in the temperature of the radiation. Yet we also know that the universe we live in today contains galaxy clusters, galaxies, stars, and planets, and even such small clumps of matter as people. How did a universe that started out so smooth get lumpy?

Recall that every particle of matter in the universe attracts every other by means of gravitational attraction. The closer to one another the particles are, the stronger they feel each other's gravitational pull. Wheeler suggests that we should think of the universe as a giant democracy in which every particle has a vote to cast in the form of gravitational attraction. A single particle has very little voting power. Only when particles band together in a voting bloc – the earth, for instance – do they manage to wield substantial gravitational clout. If we imagine a situation where all particles of matter in the universe are equidistant, with no areas in which even a few particles have drawn together more densely to form the loosest sort of voting bloc, then in that situation every particle will feel equal pull from every direction and won't budge to move closer to any other particle.

It looked for a while as though we had discovered this sort of gridlock in the super-smooth early universe – a gridlock where matter was distributed so evenly that it would never yield to form the universe we have today. That sounds like a highly unlikely situation, but if it weren't the case in the early universe, why weren't we finding even the tiniest fluctuations in the background radiation – our 'picture' of how matter was distributed in an era not long after the Big Bang? You can see that it would take only a minuscule variation in that smoothness to let gravity go to work and pull things one direction or another in ways that would show up in the background radiation. The smoothness of that radiation showed us there was a missing link in the history that would connect our contemporary universe with the Big Bang.

When George Smoot, an astrophysicist at Lawrence Berkeley Laboratory and the University of California at Berkeley, and his cohorts at several other institutions announced in April 1992 that new data from the Cosmic Background Explorer (COBE) satellite had revealed wrinkles in the fabric of the universe, wrinkles that must have been created by the Big Bang itself and not evolved later, the *New York Times* headline read: 'Astronomers Detect Proof of Big Bang.' So, in a sense, they had. The wrinkles were the fluctuations in the cosmic background radiation which astrophysicists had been looking for in vain almost from the time of Penzias and Wilson's initial discovery of that radiation in the 1960s. They were fluctuations of no more than a hundred-thousandth of a degree, but enough, the discoverers felt, to explain what had happened to the universe. These tiny variations in the topography of the universe when it was 300,000 years old were sufficient evidence of a gravitational situation in which matter would attract matter into larger and larger clumps.

There are other mysteries that those who study the Big Bang have yet to unravel. One of them has to do with the uniformity of the large-scale structure of the universe. We'll discuss that and the inflationary universe theories which may solve it in Chapter 5 in another context. Nevertheless, a wealth of evidence points to the fact that we do indeed live in a Big Bang universe.

Does an expanding universe, even a Big Bang universe, necessarily have to be a universe with a singularity at its beginning? Hawking and Penrose's calculations had said it did, but they and their colleagues were not happy with this conclusion. The singularity was derived from theory, not observation or experiment. It is a prediction we have no way of confirming or denying from observational evidence with our present technology, and perhaps not with any technology we will ever be able to invent. The theorists had discovered this Gordian knot, so it was the theorists who went to work trying to untie it. They decided to look at the origin of the universe not only with the spectacles of relativity theory, which predicts the singularity, but with the spectacles of quantum mechanics, which may not allow it.

When we study the orbits in the solar system, we're able to measure a planet's position and momentum simultaneously and get a fairly precise measurement of both. This allows us to make predictions about where the planet will be found at a later time and where it would have been found at an earlier time. We can do nothing of the sort when it comes to studying an electron orbiting the nucleus of an atom. As we've seen, one of the frustrations of quantum mechanics is that it's impossible to measure a particle's position and its momentum simultaneously and get a precise measurement for both. We don't find an electron orbiting the nucleus in the predictable way a planet orbits the sun. Quantum mechanics predicts that the probability of finding the electron is spread out over some region around the nucleus. In an article Hawking wrote in 1989, he expressed the hope – one he'd been harbouring at least since the early seventies – that in a theory of quantum gravity (a theory combining general relativity and quantum mechanics) we would find that singularities are also 'smeared out'.

As Hawking writes, 'There was a problem [in the early years of this century] with the structure of the atom, which was supposed to consist of a number of electrons orbiting around the central nucleus, like the planets around the Sun. The previous classical theory predicted that each electron would radiate light waves because of its motion. The waves would carry away energy and so would cause the electrons to spiral inward until they collided with the nucleus.' Obviously something was wrong with this prediction, because atoms don't collapse in this manner. Hawking continues:

> However, such behaviour is not allowed by quantum mechanics because it would violate the uncertainty principle; if an electron were to sit on the nucleus, it would have both a definite position and a definite velocity. Instead, quantum mechanics predicts that the electron does not have a definite position but that the probability of finding it is spread out over some region around the nucleus.
>
> The prediction of classical theory [that the electron must collide with the nucleus] is rather similar to the prediction of classical general relativity that there should be a Big Bang singularity of infinite density. Thus one might hope that if one was able to combine general relativity and quantum mechanics into a theory of quantum gravity one would find that the singularities of gravitational collapse or expansion were smeared out like in the case of the collapse of the atom.[8]

Hawking first applied this idea to the singularities in black holes, and then to the Big Bang singularity.

Hawking's theories put immense faith in the interpretation of quantum mechanics which sees the uncertainty principle as a limit upon what actually can happen in the universe, not merely a limit upon what we can measure. If we are to follow Hawking's logic, we must join him in assuming that what we cannot measure – in other words, a result at which we are incapable of arriving – cannot occur. The vast majority of physicists today are of the same mind as Hawking. Though it's not at all clear that we can apply quantum theory to the whole universe, it is possible to argue that we may need no other theory to erase the singularity, that finding everything at the same point, infinitely dense, would be simply too precise a measurement of position and momentum. The singularity is 'smeared out'. However, Hawking, with Jim Hartle, has proposed something a little more complicated than that. They and other theorists have attempted to find not only ways of ridding us of the slammed door of the singularity, but also answers to the questions which the singularity made unanswerable.

THE MAGIC OF IMAGINARY TIME

'Physicists today are not modest,' wrote physicist and astronomer Alan Lightman in *A Modern Day Yankee in a Connecticut Court*.[9] He recalls attending a lecture given by Hawking in 1984 at Harvard, where Lightman is a professor. This was shortly before Hawking had his vocal cords removed in an operation to save his life when he was suffering from pneumonia, and he could still talk in what sounded to most in the audience like low whines and moans. A student translated these sounds into words. The first shock when listening to Hawking, even with his more recent high-tech computer voice, is to find that this unlikely figure is saying anything coherent at all. The second is the supreme, understated confidence with which he ventures where others do not.

In the lecture that Lightman heard, Hawking was speaking about initial conditions – not, on the face of it, a startling subject. In an experiment, 'initial conditions' means the lie of the land at the beginning of the experiment. But, as Lightman wrote, 'I gradually realized what I was hearing: Hawking had traveled back the whole distance. For the first time, a pre-eminent scientist was tackling the INITIAL condition of the UNIVERSE – not a split second after the Big Bang, as I'd heard about before, but the very beginning, the instant of creation, the pristine pattern of matter and energy that would later form atoms and galaxies and planets.'[10]

In *A Brief History of Time* Hawking tells of a conference he attended at the Vatican in 1981. Addressing the conference, the Pope had this to say about the search for an explanation of the beginning of the universe: 'Science cannot solve such a question by itself: this human knowledge must raise itself above science and astrophysics and what is called metaphysics; the knowledge must come above all from the revelation of God.'[11] Hawking, of course, would have none of that – though to describe this statement, as he does in *A Brief History of Time*, as a 'prohibition' against the search for the beginning of the universe seems an overreaction.

At the same conference, Hawking presented a proposal that there was no beginning of the sort the Pope was speaking of – a proposal that there were no boundaries for the universe. Hawking had decided that that holy of holies, the singularity, might not be a block to our knowledge after all. In order to arrive at this proposal, he and Hartle used the device of imaginary time.

Imaginary numbers, contrary to popular legend, were not invented by Hawking but have been around since the mid-sixteenth century. They deserve some demystification. They are a mathematical, not a metaphysical, concept, despite some early ruminations which might suggest the contrary. Gottfried Leibniz, the seventeenth-century mathematician who narrowly, and perhaps unfairly, lost the race with Newton to claim to be the inventor of calculus, saw imaginary numbers as a 'sort of amphibian, half-way between existence and non-existence'. He suggested that they were somewhat like 'the Holy Ghost in Christian theology'.[12] However, there is nothing mystical in the least about imaginary numbers.

Imaginary numbers are not even a very complicated mathematical concept, although the way Hawking and Hartle have applied them to the universe is not easy to understand. They are numbers which when squared yield a negative number. If you never went beyond the more elementary maths courses in school, you

Figure 4.1

Romancing the creation

© *The Washington Post Writers Group, reprinted with permission*

Berke Breathed

probably didn't encounter them. You were taught that the square of -4 is 16, just as is the square of 4. The square of any number, negative or positive, is a positive number. If this is true then you can't possibily ask what is the square root of -16. The situation is different with an imaginary number. The square of imaginary 4 is -16. Imaginary 3 squared is -9. The square root of -16 is imaginary 4; the square root of -9 is imaginary 3.

What then is imaginary time?

According to Big Bang theory, in the very early universe, space was extremely compressed. Here, Hawking suggests, the smearing effect of the uncertainty principle could erase a basic distinction, which still endures in relativity theory, between space and time dimensions. Allowing the time coordinate to be an imaginary number provides a new way of looking at this situation in which it is more accurate to talk not about three dimensions of space and one of time but, instead, of four dimensions of space. Time, in this approach, becomes indistinguishable from a space dimension. To quote Hawking, 'Calculations suggest that this state of affairs cannot be avoided when one considers the geometry of the universe during the first minute fraction of a second.'[13]

The idea of treating time as a space dimension is not new to physics. Physicists use this device for working out problems in quantum mechanics. What makes Hartle and Hawking's a radical approach is that they don't merely use this trick to solve a problem and then go back to the usual concept of time. They propose that time really was like space. As Hawking has said, 'I think these concepts will come to seem as natural to the next generation as the idea that the world is round. Imaginary time is already a commonplace of science fiction. But it is more than science fiction or a mathematical trick. It is something that shapes the universe we live in.'[14] We can't simply accept this statement from Hawking, or say 'Time really was like space', without recognizing that in doing so we leap-frog a great deal of discussion about the reality of mathematical models and about reality itself. But let us proceed for the moment and return later to quibble about that.

If Hartle and Hawking's proposal is correct, we don't have to worry about time and space beginning in a singularity, because here, the tiniest interval away from what we have been assuming was the beginning, in imaginary time it becomes meaningless to talk about 'past' at all. The concept of a beginning 'before' that is also meaningless.

The question then remains as to the geometry of this four-dimensional space. It has to join smoothly onto our familiar space-time as the universe expands and the quantum smearing effects subside. One possibility among many others – an infinite number of possibilities, says physicist and science writer Paul Davies, echoing Poincaré – is that the four-dimensional space curves around to form a closed surface, without any edge or boundary at all, a situation similar to our earth or to the balloon on which our imagined ant lived, but with more dimensions. The ant, you remember, found no boundary or edge. There are no boundaries to Hartle and Hawking's universe, no boundaries in space and – far more significantly – no boundaries in time. No beginning. The concept of 'past' ends in the early universe just as the concept of 'north' ends at the north pole, without a boundary or an edge off which to fall – without a beginning (Figure 4.2). What can we say then about 'initial conditions'? As Hawking puts it, 'The boundary conditions of the universe are that there are no boundaries.'[15]

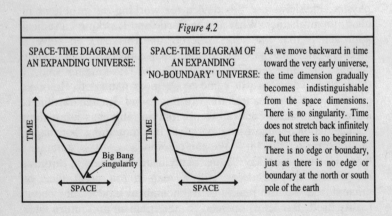

Figure 4.2

| SPACE-TIME DIAGRAM OF AN EXPANDING UNIVERSE: | SPACE-TIME DIAGRAM OF AN EXPANDING 'NO-BOUNDARY' UNIVERSE: | As we move backward in time toward the very early universe, the time dimension gradually becomes indistinguishable from the space dimensions. There is no singularity. Time does not stretch back infinitely far, but there is no beginning. There is no edge or boundary, just as there is no edge or boundary at the north or south pole of the earth |

'No boundaries' might seem to imply 'infinite', but in fact it doesn't. In the case of the surface of the earth, there are no boundaries in space, and yet the surface of the earth is not infinite in size. So it is with Hartle and Hawking's no-boundary universe. Space is not infinite, nor is time.

Hartle and Hawking prefer this geometry for reasons of mathematical elegance. What possible reason do you and I have for believing with them that this proposal could represent physical reality, that time really might have been like space, and that this scenario is not merely a mathematical fiction or an article of faith arising from a yearning for mathematical beauty and an explanation of the universe which doesn't require a *deus ex machina*? The question is not only 'Could it really have happened this way?' but also 'If it could have, why should we think it did?'

Hawking is the first to point out that his idea is just a proposal. He doesn't even call it a theory. It's a spectacularly wild leap of imagination. He hasn't deduced these boundary conditions from some other principle. Of course it goes almost without saying that we have no direct observational data, but, having made the leap, Hawking and others have carried matters forward by asking what sort of universe would result from this particular 'no-boundary' situation. The calculations are extremely complex, and so far they've been carried out only in simple models, but they seem to demonstrate that the proposal can be linked by mathematical consistency to the real universe as we observe and experience it, that the universe that would result would indeed be a universe like our own. In real time, where we live, it would still appear that there were singularities in black holes and at the beginning of the universe. But in imaginary time there would be none in either place.

This isn't then just the Land of Oz. So far so good. However, mathematical and logical consistency do not demand this model of the universe as opposed to others. Nothing has so far shown that it is the only consistent model or one to be strongly preferred over others.

Could it have happened this way? It's far too early in the game to answer that question. Did it happen this way? Only on aesthetic and philosophical grounds, and because it upholds one of the assumptions of science, is it possible at present to prefer this theory over others. Hawking tells us that the proposal appeals to him because 'it really underlies science . . . it is really the statement that the laws of science hold everywhere.'[16] It is that – a statement, not a demonstration that they do or that

Hartle and Hawking have correctly described the manner in which they do.

Imaginary time also plays a large role in theories from Hawking and others about wormholes and baby universes, perhaps an even more spectacular leap of imagination, though in this case the concept arises from previous ideas, particularly those of Wheeler.

Once more, picture a balloon – an enormous one – inflating rapidly. This is the cosmic balloon, our universe. Picture also dots on the balloon's surface to represent stars and galaxies, and picture them causing tiny dimples and puckers in the surface. These are the curving of spacetime caused by massive objects, which Einstein predicted. Imagine also that, in spite of these little puckers, the surface is relatively smooth, even when we look at it through a not very powerful microscope. If we look at it through a much more powerful microscope, we find it isn't smooth after all. The surface seems to be vibrating furiously, creating a blur, a fuzziness.

We've encountered such fuzziness before. The uncertainty principle causes the universe to be a blurry affair at the quantum level. The surface of the cosmic balloon is uncertain in a similar way. Under high enough magnification the quantum fluctuation becomes such that Hawking claims there's a probability we'll find it doing – as he puts it – *anything*. Specifically, he thinks there's a probability that the cosmic balloon will develop a little bulge in it. On rare occasions you see this happen as a party balloon is inflated. On even rarer occasions the bulge doesn't cause the balloon to burst, but instead turns into a miniature balloon attached to the surface of the larger balloon by a narrow neck. If you saw this happening with the cosmic balloon, you'd be witnessing the birth of a baby universe. The little neck would be a 'wormhole'.

Is there data to support this supposition? Surprisingly, that isn't such a ridiculous question when it comes to wormholes and baby universes, although it won't be direct observational data. Several experiments have been proposed. However, Hawking himself doesn't think these tests will succeed in determining whether or not wormholes and baby universes exist. When we speak about seeing all this through a microscope we're being fanciful. If anything can be said to start small, it's a universe. The most probable size for a wormhole connection between our universe and the new baby is only about 10^{-33} centimetres across. If you want to write that out as a fraction you do so by using 1 as the numerator and 1 followed by thirty-three zeros for the denominator. A wormhole is like a tiny black hole, flickering into existence and then vanishing after an interval too short to imagine. Another reason why we can't witness the birth is that it all happens in imaginary time.

However, the baby universe attached to this wormhole umbilical cord may not be so short-lived. Nor must it necessarily continue to exist only in imaginary time. Eventually the new universe may expand to become something like our present universe, extending billions of light years. Perhaps not only something like ours, but exactly like ours, with galaxies, stars, planets, life. In fact the suggestion Hawking makes is that our universe did originate that way, as a baby universe bulging off the side of another universe. According to the theory, there may be many universes, a never-ending labyrinth of them, connected by wormholes in more than one place. There might even be wormholes connecting one part of our universe with another part, which would allow for rapid travel between very distant locations – travel in space or even in time – if we were small enough and if we could travel in imaginary time (see Figure 4.3). It does seem that among elementary particles it's not completely unreasonable to quote e. e. cummings:

> Listen, there's a hell of a
> good universe next door:
> let's go![17]

Wormhole theory not only proposes to rid us of the problem of singularities and explain another way the universe may have begun, it also attempts to solve a puzzle we call the cosmological constant problem. We'll save that for Chapter 5. Meanwhile, the no-boundary proposal and wormhole/baby-universe theory are not the only suggestions for unravelling the Gordian knot of the singularity.

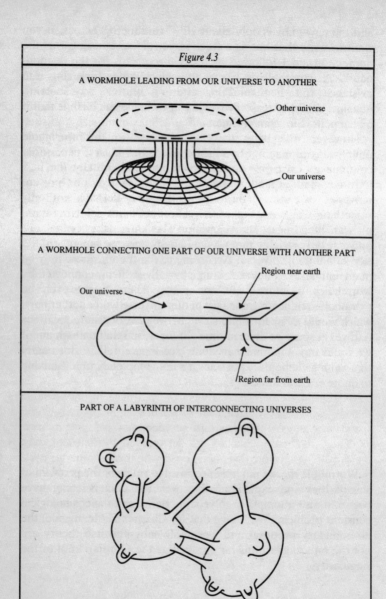

Figure 4.3

A WORMHOLE LEADING FROM OUR UNIVERSE TO ANOTHER

Other universe

Our universe

A WORMHOLE CONNECTING ONE PART OF OUR UNIVERSE WITH ANOTHER PART

Region near earth

Our universe

Region far from earth

PART OF A LABYRINTH OF INTERCONNECTING UNIVERSES

Drawing by Andrew Dunn

116

THE PULSING UNIVERSE AND THE ELUSIVE
CLUE OF DARK MATTER

Will there come a time when the universe stops expanding and begins to contract again? Friedmann's solutions to Einstein's equations suggested the possibility of three types of universe. In one model, the universe expands to a maximum size and then recollapses. In a second model, the universe expands rapidly and never stops expanding. In a third model, the universe expands at exactly a critical rate to avoid recollapse (see Figure 4.4).

How can we find out which model fits our universe? To do so, we have to compare the current rate of expansion with the current average density of mass in the universe. There are problems with measuring the current average density of mass.

The amount of gravitational attraction between objects depends upon the amount of their mass and their distance from one another. We ought to be able to add up the voting power (as Wheeler thinks of it) of all the matter in the universe and be able to calculate at least roughly whether that will produce enough gravitational attraction to stop the expansion and close the universe. When we do this calculation, we find that the amount of matter in the universe that we can observe directly isn't nearly sufficient to stop the expansion. You might expect the discussion to end there, but it doesn't.

We have good reason to suspect there is mass in the universe that we can't observe because it isn't radiating in any part of the spectrum – hence its name, 'dark matter'. First, we have indirect evidence that many galaxies are surrounded by halos of dark matter. We determine this not by observing the dark matter itself but by observing the movement of visible matter such as stars and gases within the galaxy. For instance, the mass and distribution of visible matter in our own galaxy is not enough to account for the way our galaxy rotates. The rotation indicates that the mass of the halo must be much larger than the visible mass of the galaxy.

In June of 1993 Douglas Lin of the University of California announced that study of the orbit of the Large Magellanic Cloud, a satellite galaxy of the Milky Way, provides additional evidence that a dark-matter halo surrounds our galaxy. His calculations indicate that the Milky Way galaxy must weigh 600 billion solar masses (600 billion times the mass of our sun). That is five to ten

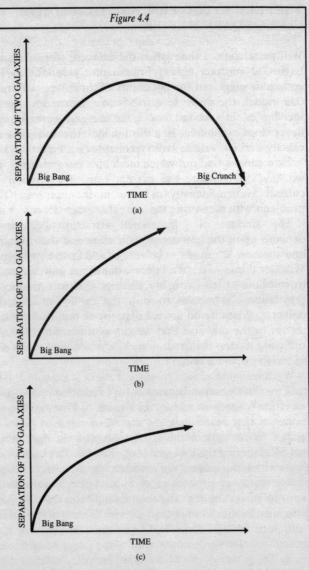

Figure 4.4

Three models of the universe: (a) The universe expands to a maximum size and then recollapses. (b) The universe expands rapidly and never stops expanding. (c) The universe expands at exactly a critical rate to avoid recollapse

times the mass of all the visible material in our galaxy. The visible part of the Milky Way is about 120,000 light years in diameter. The total diameter of the visible galaxy and the dark-matter halo might be 800,000 light years in diameter or more.

There are other observational clues. Observations of distant stars and galaxies show us effects that can best be explained as lensing effects, where light from these remote sources is bent by massive objects or accumulations of mass nearer to us – in the way the sun bends the paths of light from distant stars. By studying these effects, astronomers are able to calculate how much mass is there, even though the mass is invisible. These studies continue, and obviously will take some time. There is a great deal of universe out there. Until we can find out more precisely how much dark matter exists, we can't determine which Friedmann model of the universe is correct. The case is not closed, and neither, necessarily, is the universe.

However, let us suppose for the moment that the Friedmann model in which the universe will some day contract again is the correct one to describe our universe. What's to keep it from expanding again after it contracts? Indications coming from quantum physics and supersymmetric string theory are that the universe wouldn't necessarily contract all the way to a singularity. Instead, just short of that, it might 'bounce' and start the cycle all over again. How would we know whether this model of a 'pulsing' universe is correct?

First, of course, the universe can't bounce or pulse if it doesn't fit the Friedmann model in which it contracts. A second consideration has to do with entropy.

To measure the amount of entropy in a system means to measure the amount of disorder. The Second Law of Thermodynamics says that entropy (disorder) in any closed system cannot decrease, it can only increase. This law can in rare instances be broken, and we will see in Chapter 6 that there are theories which call into question the universality of the trend toward disorder; but it is generally accepted that entropy in the universe as a whole is inexorably on the increase.

This may seem to defy common sense. Obviously if we have marbles of two colours in a box separated by a partition, with all the reds on one side and all the greens on the other, and if we remove the partition and shake the box, there is only the tiniest of probabilities that at a future time the marbles will again sort themselves out by colour as they were before the partition was removed. However, all we have to do is reach into the box ourselves and re-sort the marbles. Haven't we defeated the Second Law of Thermodynamics? No – we haven't. Our reaching into the box means it isn't a closed system. Similarly, we can put some bit of the universe in order, perhaps wash the dishes, stack them neatly on the shelf, and sort the garbage and recyclables, but the bad news is that in the physical and mental effort of doing all this we convert energy to a less useful form and this adds to the overall entropy of the universe. You can combat entropy by never doing anything at all, but merely staying alive converts some energy.

A way to understand this situation is to consider the fact that in any system, the start-up conditions which would allow things to progress from disorder to order are vastly more rare than the start-up conditions which would allow them to progress from order to disorder. For example, all the marbles in the box would have to be rolling at precisely the right speeds and in precisely the right directions to get back to their sorted positions on the two sides of the box. For that to happen is not impossible, but it is far from likely in view of all the other speeds and directions that would not get them there.

This Second Law of Thermodynamics is one of the great organizing principles (though perhaps it would seem more appropriate to call it a disorganizing principle) of the universe, and it appears to have a great deal to do with our distinguishing the past from the future. Remember those kindergarten exercises in which someone asked you to take four pictures and put them in order? Even at age three or four you knew that the one showing the bull entering the china shop doorway with all the china immaculately displayed in the showcases was most likely going to be picture number one, not picture number four.

Why should entropy cause a problem with the model of a universe that pulsates? The problem is that when one cycle of expansion and collapse is finished, the universe must surely be at a much higher level of entropy or disorder than it was at the beginning of that cycle. Penrose, whose work with Hawking, you will recall, led to the theoretical confirmation of the Big Bang singularity, insists that entropy at the beginning of the universe

Figure 4.5

From The Arrow of Time, by Peter Coveney and Roger Highfield, W.H. Allen (Virgin Publishing Ltd)

(A)

(B)

The proverbial bull in the china shop. We never see the time-reversed process in which B would occur before A and entropy at the end would be 'ridiculously different'.[18] The universe at the Big Bang is so highly organized that if it were cut in half it would show almost no structure. The universe at the Big Crunch will be a great mess. The upshot is that, unless there were some as yet unexplained way of putting things back in order very quickly before the next expansion, the next expansion would begin with a much higher state of entropy and would produce a different sort of universe. We may be living in the only recurrence of the cycle in which it would be feasible for us to live.

There's another possibility. Perhaps when the universe reverses itself and contracts, the arrow of entropy also reverses itself. Perhaps in a contracting universe entropy decreases, broken teacups reassemble, bulls run tail-first through china shops leaving once-shattered china sitting whole upon the shelves. A further implication might be that the universe would reverse itself not only in space dimensions, but also in the time dimension. We can imagine a science-fiction-like scenario in which everything that had happened in the expanding phase would happen backward in the contracting phase. The cycles would be endless repetitions. If that's so, I'm not sure I want to know about it – nor would I, according to astronomer Thomas Gold, who first proposed the idea. He suggests that intelligent beings might have their thought processes reversed in the contracting phase so they would not notice the difference. They would still see themselves as living in the expanding stage.

Both Hawking and Penrose think that the arrow of entropy would not change in the contracting phase. Entropy would continue to increase. If they're correct, then some calculations indicate that the cycles of expansion of a pulsing universe would get bigger and bigger, and endure longer and longer, and that there would be no end to this process. Other calculations suggest a different picture: a pulsing universe might not be any more eternal than the successive bounces of a rubber ball, gradually running down. Although with the model of a pulsing universe we may have circumvented our own particular singularity, we haven't necessarily erased the notion that somewhere, perhaps several pulses back, there might have been a beginning that is still waiting to be explained.

It's possible that a universe that did not contract again might also be a cyclical universe.

THE MYSTERIOUS WOBBLING
OF NOTHINGNESS

It was Alan Guth of the Massachusetts Institute of Technology who applied the attractive phrase 'free lunch' to the universe. Like Wheeler, Guth has a reputation for thinking up catchy names. His comment was: 'I have often heard it said that there is no such thing as a free lunch. It now appears that the universe itself is a free lunch.'[19]

The idea predated Guth's christening of it. American physicist Edward Tryon proposed in 1973 that quantum mechanics and relativity, fused in a quantum theory of gravity, might show us a mechanism for creating the universe out of nothing – *ex nihilo*. Beginning in 1978, cosmologists at the Free University of Brussels provided a series of suggestions along those lines. The idea originated as a way of explaining the creation of matter, and only later led to something more fundamental, an explanation for the creation of spacetime itself. Let us first see how it might apply to the creation of matter.

Suppose it all began with a vacuum where spacetime was empty and flat. The uncertainty principle doesn't allow an emptiness of complete zero. We've seen earlier that it rules out the possibility of measuring simultaneously the precise momentum and the precise position of a particle. It also rules out other simultaneous measurements. The one that concerns us here has to do with fields, such as a gravitational field or an electromagnetic field. If we measure the value of a field, we can't at the same time measure precisely the rate at which that field is changing over time, and vice versa. The more precisely we try to measure the one, the fuzzier the other measurement becomes.

In complete emptiness, the two measurements would read exactly zero simultaneously – zero value, zero rate of change – both very precise measurements. The uncertainty principle doesn't allow both measurements to be that definite at the same time, and therefore, as most physicists currently interpret the uncertainty principle, zero for both values simultaneously is out of the question. Nothingness is forced to read – something.

If we can't have nothingness at the beginning of the universe, what do we have instead? A continuous fluctuation in the value of all fields, a wobbling a bit toward the positive and negative sides of zero so as not to *be* zero. But these are energy fluctuations. How do we get matter out of this process?

According to Einstein's equation $E = Mc^2$, there can be no increase of E (that is, energy) on one side of the equal sign unless there is also an increase of M (mass) on the other side. (The c is the speed of light and that can't change.) Because of this equivalence of mass and energy, a quantum energy fluctuation would produce the equivalent mass of particles. These particles would attract each other by means of gravity, causing flat spacetime to become curved.

In this scenario, it would seem that the creation of matter violates the generally accepted rule that energy or matter cannot be added to or subtracted from the universe. Some have argued that such a violation smacks of divine intervention. But we really have no violation in this case. The gravitational attraction is negative energy, which offsets the positive energy of the particle masses – leaving a net gain of zero. Thus the instability and unpredictability of the flat spacetime quantum vacuum seeds the birth of the universe.

This leaves open the possibility of another kind of cyclical universe. Suppose the universe that emerged from this process turned out to be the sort of universe that goes on expanding for ever. The matter in the universe would spread out thinner and thinner and would eventually become extremely dilute – a situation very like the flat, 'empty' spacetime with which this story began. Perhaps the entire drama would then repeat itself on a far grander scale.

Either this process has been repeated an infinite number of times in the past, or else we still need a way to explain how it began the first time. An even more basic version of this free lunch creation proposes how spacetime itself came into being. We've seen that events we observe on the quantum level can be 'uncaused events' – happenings without a certain history. Physics theorists are still in the process of trying to explain gravity in a quantum mechanical way, but some think that doing so will show us an even more fundamental uncertainty which might allow the creation of space and time to occur spontaneously, without cause. There may be a mathematically determinable probability that a snippet of spacetime would emerge from nothing at all.

We've observed such uncaused events only on the super-microscopic level, and so we assume that is the only level on which they occur, but we needn't think that just because we are applying this process to the creation of the universe we are operating on a size level larger than that studied by quantum physics. The size of the seminal bit of space would probably be the size of Hawking's wormhole, 10^{-33} centimetres. We've already seen that such a tiny speck of creation can grow to be a universe.

As the saying goes, 'Nothingness is unstable, and tends to decay into something.' Calculating the probability of there being something rather than nothing, it seems that there is more likely to be something. Thus physics attempts to update Thomas Aquinas's assertion in the thirteenth century that 'We cannot but admit the existence of some being having of itself its own necessity, and not receiving it from another, but rather causing in others their necessity. This all men speak of as God.'[20] The 'free-lunch' argument is that it may not be God which has 'of itself its own necessity', but simply a highly likely snippet of spacetime – which might also answer Hawking's question 'Why does the universe go to all the bother of existing?'[21] Because it would be considerably more bother not to exist!

If any of the proposals we have been discussing is correct, the origin of the universe is no longer beyond the laws of physics or unknowable to us. There is no slammed door – at least not just *there*. But at first glance, to those not accustomed to considering mathematics such a powerful guide to reality, these theories seem like ripping science-fiction yarns rather than science fact. We can get quite carried away reading about them. We envision the wormholes, or we imagine time swooping in to join the space dimensions, or we fancy the wobbling of nothingness and the minuscule morsel of somethingness destined to expand and be the entire universe. But then we raise our heads from the book, glance around at the four walls and the trees outside the window and perhaps a chair like my Texas chair, solid and quiet over there against the wall, and we think we have returned to reality. What claim does all this science which borders on science fiction have to being 'real' in the way the familiar objects around us seem to be 'real'? What actual relevance does any of this have to whether or not we believe there is a real God?

We hear the argument that the Big Bang supports the biblical view of creation and is a threat to atheism. We hear the argument that Hawking's no-boundary proposal abolishes the need for God. In order to support anything or be a threat to anything, a theory must have some claim to being the correct model of what really is.

We'll begin this discussion with the Big Bang theory, asking: How valid is the claim that this theory is an accurate retelling of the history of the universe?

The Big Bang was never a purely mathematical theory. It arose out of a combination of observation and theory. Though it doesn't have as firm an underpinning of observational data, and certainly not as much fruitfulness for practical technology, as relativity and quantum mechanics have, it is not a speculative theory like the no-boundary proposal. In line with the criteria we discussed in Chapter 3, Big Bang theory, far more than its erstwhile competitor the Steady State universe, accounts for a wealth of available evidence in a relatively simple, efficient, and unartificial way; and it ties in with other strong theories in such a way as to make eminent sense and suggest further meaningful lines of inquiry and thought.

The theory does still leave us with mysteries and loose ends. However, we can say that the Big Bang is currently regarded as a well-established theory, the 'standard model' acceptable to most physicists, and that the questions that remain do not cast serious suspicion on it. They are more a matter of settling which specific version of the theory is correct – shall we accept inflation theory, for example (we'll get to that in Chapter 5) – working out details, improving, and refining. What claim does the Big Bang theory have to being the real history of the universe? A good claim. What actual relevance does the theory have to whether or not we believe there is a real God? If one's atheism or agnosticism rests on the hope that the Big Bang theory is not the correct version of history and will eventually be replaced by a different model entirely, it would be best to look for other support. But it's doubtful, in spite of some earlier panic, whether anyone's atheism or agnosticism is threatened by this theory in the 1990s.

'In the beginning, God created the Heavens and the Earth.'[22]

In line with Big Bang theory (with singularity), that might more specifically read: 'In the beginning, God created everything that was later to become what we now call the Heavens and the Earth, as well as the laws that directed that outcome, and God caused it all to begin happening.' For those who accept the Genesis account as metaphoric or symbolic, or see it as a beautifully poetic but inadequate human description of events whose magnitude defies any human description – even a scientific explanation – the connection is significant. The Big Bang singularity, by slamming the door in our faces, puts us in the uncomfortable position of not being able to explain how the universe began. It doesn't necessarily follow that the unknowable explanation is God, but it would seem that God is at least as good an explanation as any other. Nevertheless, the Big Bang account does not support a word-for-word acceptance of Genesis.

There are those who believe in God who see no philosophical advantage in the Big Bang over Steady State theory. They point out that the Judaeo-Christian God creates and sustains the universe continually and perhaps eternally (if the universe is eternal), and that whether or not there was a beginning of time has no relevance for the question of whether or not God is the creator.

We must now inquire with regard to those proposals which attempt to undermine the singularity – the no-boundary proposal, wormhole and baby-universe theory, the pulsing universe theories, and the free-lunch universe: What claim do these have to being descriptions of something that really happened ten to twenty billion years ago, and what relevance do they have for whether or not we believe there is a God?

These proposals were not developed in direct response to observational data, and they have so far no direct experimental or observational data to support them. It is correct to say that some things we have been able to observe *suggest* . . . but not to say we have direct supporting evidence. We've detected what seem to be uncaused events in observation of the quantum level, but it is not yet clear that we can apply what we know about quantum mechanics to the entire universe. In any case, observing the quantum level in the way we are capable of observing it is not the same as looking at the universe when it was 10^{-35} seconds old or even younger. Before reaching the temperatures and densities of that era, we run out of physics which has been tested in any laboratory.

These proposals began as flights of fancy, though some of them have become more than that. Their claims to being correct rest primarily upon arguments of mathematical and logical consistency and the elegance of that consistency. However, it has not been established to everyone's satisfaction that Hartle and Hawking's no-boundary proposal is indeed internally consistent. Whether it is consistent with well-accepted, well-established knowledge about the universe, whether calculations and simulations based on the theory produce a universe like our own, and whether it is consistent with other speculative but highly regarded theories such as superstring and inflation theory – these are questions which have been answered only in a very preliminary way. Superstring theories are at present the favoured candidates for unifying the forces of nature – though string theory is arguably as difficult to verify through experiment and observation as the origin-of-the-universe proposals are. There are still many versions of string theory, and it is difficult to decide with which version compatibility would be meaningful.

If we claim we are approaching an ultimate theory of the universe, we must remember that the closer we get to such a theory the more significant the question becomes: Is this the ONE theory which succeeds in being mathematically and logically self-consistent, encompassing all the data and all approximate theories, explaining constants of nature, and producing a universe like our own, while all other theories fail to do so?

We are not even remotely near establishing that any present theory is unique in these ways. The proposals we've seen concentrate on initial conditions. Only in combination with other theories (superstring theory, perhaps) might they approach anything like Theory of Everything status. But even when it comes to describing initial conditions, no-one has been able to show, with any of these proposals, that mathematical consistency looks likely to constrain us to this model and this model alone. Davies has pointed out that there are infinite possibilities for the geometry of four-dimensional space in the early universe. Hartle and Hawking picked one geometry over the others because of its mathematical elegance. But they have not eliminated other possibilities by showing that theirs is the only geometry that is mathematically and logically consistent.

If the success of one theory depends in such great part on the failure of competing theories, then there must be some way competing theories can fail – which brings us to the question of falsifiability. None of these proposals is at present falsifiable by direct experiment or direct observation. Their falsifiability lies primarily in the possibility of finding flaws in the internal mathematical logic, discovering that the proposal is incompatible with other more well-established theory, or showing that the model is incompatible with the universe as it has actually evolved – that is, showing that you can't start the universe as the theory proposes and have it eventually turn out to be the universe we know today.

If evaluation of these theories must rely heavily upon mathematical consistency, it behoves us to ask whether we are willing to think of mathematics as so infallible a guide. We saw, in Chapter 3, Barrow's point that mathematics is not in all cases self-consistent but is capable of producing contradictory solutions. He goes on to say that it seems not possible to discover these inconsistencies except by accident. We cannot go about systematically finding out where they lurk and how to avoid them. They may lie hidden in the mathematics that underpins many modern-day physics theories. We are not being incorrigible sceptics to wonder whether we can arrive at any reliable conclusions about the real universe by means of mathematics alone.

REALITY IN THE ABSENCE OF APPLES

When Hawking wrote that a mathematical theory 'exists only in our minds and does not have any other reality (whatever that may mean)',[23] he was not simply being lazy about defining terms. What 'reality' means is precarious in anyone's language. In the language of scientific theory, defining 'reality' becomes even more complicated. There is mathematical reality in the sense of mathematical logic and consistency, but does that reality necessarily translate to reality as we know it on the common-sense level, or to reality on the ultimate chair-as-it-is-in-itself level? We know that maths says that $2 + 2 = 4$ and we can see that having two apples and adding two more will indeed give us four apples, and that is 'real' in common-sense terms. But what reality does this actually allow us to assign to the equation $2 + 2 = 4$ in the absence of apples and all other objects we can count?

Although most of us don't think of ourselves as espousing one philosophy of mathematics or another, to a certain extent each of us does – not, admittedly, a philosophy that is often consciously or rigorously worked out or one we advertise on our bumper stickers. To a surprising degree, your attitude or mine toward a mathematical theory as a guide to the history of the universe, and even how that attitude may affect our personal religious beliefs, depends on our philosophy of mathematics. A short perusal of the philosophical possibilities on offer is clearly to be recommended.

Most of us, when we first learned mathematics in school, probably assumed it was something that had been invented by humans. Mathematics was a way people had devised to make sense of things, put them in order, and keep track of them – a brilliant system, improving all the time as mathematicians worked on it. But mathematics wouldn't have existed if human beings hadn't existed.

If we are correct to think of mathematics as a human invention, then we are on shaky ground to assume that it will always and in all circumstances allow us to predict what physical reality will be like. We must go on discovering physical reality and inventing mathematics to describe it, and avoid the temptation to use what may be inappropriate mathematics to predict far ahead of discovery.

I remember clearly when it first dawned on me that human beings might have discovered mathematics, not invented it; that it might lie waiting in nature; that mathematical truth might be a part of independent reality. It wasn't in mathematics class, but in music theory, when I studied the harmonic series. It seemed to me that this pattern could not be a human way of sorting things out. It would have existed even if human beings had never existed. If I was right about mathematics being inherent in nature, then human mathematics could be successful only insofar as it accurately reflects the situation which is already there in nature. I didn't take this to mean that mathematics as we know it

130

actually does adequately capture reality. But if nature is inherently mathematical, that did seem to imply that *some* fundamental form of mathematics, as we know it or have yet to discover it, does. The concept of mathematical truth being transcendent, objective truth is expressed by Penrose in his book *The Emperor's New Mind:*

> How 'real' are the objects of the mathematician's world? . . . Can they be other than mere arbitrary constructions of the human mind? . . . There often does appear to be some profound reality about these mathematical concepts, going quite beyond the mental deliberations of any particular mathematician. It is as though human thought is, instead, being guided towards some eternal external truth – a truth which has a reality of its own, and which is revealed only partially to any one of us.[24]

The philosophy which sees mathematics as inherent in nature, rather than invented by human beings, is compatible with thinking God is First Cause of the universe, in the sense summed up by Aquinas – 'having of itself its own necessity, and not receiving it from another, but rather causing in others their necessity'. God would then be the divine inventor of mathematical truth.

However, it is this philosophy of mathematics, as discovered, not invented by humans, which also allows us to consider mathematical and logical consistency as a more powerful concept which God had no choice but to adhere to in creation. It even allows us to consider mathematical and logical consistency as a strong contestant for First Cause, not only constraining the universe to be what it is but making its very existence inevitable. Is it perhaps mathematically and logically *in*consistent for the universe *not* to exist precisely as it does? The answer to Hawking's question 'What is it that breathes fire into the equations and makes a universe for them to describe?'[25] might be that the equations are the fire.

An even more extreme form of the philosophy which sees mathematics as ultimate, objective truth is to believe that existence as a mathematical model IS reality. Maybe the equations aren't just the fire. Maybe they are the universe. As

Barrow explains this point of view, 'Life *must* exist in every sense because there exists a mathematical model of it.'[26]

We'll round out our list of philosophies of mathematics by mentioning two more. Some see mathematics as nothing more nor less than a system of logical deductions and connections, a great network of self-consistency, making it something like a game and side-stepping questions about its meaning or reality. However, Kurt Gödel's incompleteness theorem (that in any mathematical system rich enough to include the addition and multiplication of whole numbers, there must exist mathematical statements whose truth or falseness can't be decided from within the system) showed that mathematics can never be bundled up in any such neat, self-contained package.

A fourth philosophy confines mathematics to sequences of step-by-step logical constructions, much the way a computer operates. There was a time when we thought computers would be able to carry out all mathematical operations, but we now know that mathematics contains non-computable functions. This fourth way of looking at mathematics has nothing to say about whether, when functions of mathematics are non-computable, they maintain any practical link with reality. We know that there are mathematical operations which can't be simulated by a computer program, and we need some of these operations to understand the physical universe.

In the light of these four interpretations, it is interesting to find Hawking dodging the issue of reality in the following statement:

> If you take a positivist position, as I do, questions about reality do not have any meaning. All one can ask is whether imaginary time is *useful* in formulating mathematical models that describe what we observe. This it certainly is. Indeed, one could even take the extreme position and say that imaginary time was really the fundamental concept in which the mathematical model should be formulated. Ordinary time would be a derived concept that we invent as part of a mathematical model to describe our subjective impressions of the universe.[27]

In other words, ordinary time is a partial or approximate description which is useful for coping with common-sense experience, while imaginary time may be a more fundamental description, useful to explain the universe. Hawking prefers to avoid questions about what is real. To his way of thinking, discussions about things we can never know – such as the question of which kind of time is more 'real' or whether there is a God – are not 'useful' and cannot possibly be relevant to a decision about reality. Perhaps he is right.

However, when we adopt that way of thinking, we run a risk of redefining reality rather than avoiding discussion of it. We fall into a habit of adopting 'what is useful' and 'what we can know' as our new 'reality'. It is this risk that Hawking's wife Jane was referring to when she told an interviewer in 1988, 'There's one aspect of his thought that I find increasingly upsetting and difficult to live with. It's the feeling that, because everything is reduced to a rational, mathematical formula, that must be the truth.'[28] The suspicion that we may end up with a straitjacketed and distorted picture of reality if we cling unwaveringly to a belief that truth is intrinsically mathematical has been shared by some of our greatest physicists and mathematicians, among them Ludwig Boltzmann and James Clerk Maxwell.

Nevertheless, mathematical consistency and beauty are an exceptionally effective pointer in science. We know so not from philosophy or as an article of faith, but from long experience. As Davies wrote in his book *The Mind of God*, 'much of the mathematics that is so spectacularly effective in physical theory was worked out as an abstract exercise by pure mathematicians long before it was applied to the real world . . . and yet we discover, often years afterward, that nature is playing by the very same mathematical rules that these pure mathematicians have already formulated.'[29] Whether it necessarily follows that nature in all contexts, even at the split second of its origin, played and will continue to play by those rules, we don't of course know. If mathematical truth is discovered, not invented, that would seem to give us greater cause for confidence, but even so we can't assume we're reading nature's mathematical rules aright, and aren't merely over-confidently projecting known rules upon regimes where they no longer apply, while failing to account for nature's deeper mathematical reality.

Theoretical physicists, however strong their belief in mathematics, do feel obliged to show the connections between their mathematical theories and the world most of us more readily regard as 'real'. Until those connections are clear, no-one, including the theorists, pretends that any of these proposals we've been discussing are 'scientific knowledge' in the way relativity theory or quantum mechanics is.

In addition to their mathematical beauty and consistency, what makes these origin-of-the-universe proposals particularly attractive is their ability to circumvent the singularity, and this is a more dubious argument. Listening to Hawking, it sometimes seems that the strongest support for his no-boundary proposal lies in the fact that it upholds the assumptions of science that there are laws of physics which apply everywhere and that it is not beyond human capacity to discover what they are. As dearly as we may hold those assumptions and as well as they've served us in the past, when it comes to arguing for the validity of a proposal for the origin of the universe, these are self-serving arguments – good arguments maybe for hoping a theory is correct, but no arguments for deciding it is. Such a decision would be an act of faith.

WHAT PLACE FOR A CREATOR?

John Polkinghorne, Cambridge theoretical physicist and theologian, wrote in *The Cambridge Review:*

> Those who essay a quantum cosmology are necessarily skating on intellectual thin ice, however pretty the arabesques they perform. Needless to say, Stephen Hawking is well aware of this problem. He believes that sufficient of the lineaments of an eventual theory of quantum gravity can be discerned to make at least general sense of the cosmological programme. Doubtless Steve's speculations deserve to be taken more seriously than those of many other practitioners, but they remain speculations nevertheless.[30]

To say that Hawking or any other theorist has shown us there is no God is premature to say the least. Nevertheless, that doesn't end the discussion of the relevance of these theories for religious belief. They exert great power over our thinking about God and the universe. Why, if they are so unproven, such acts of faith themselves? Because they undermine one reason for believing that there is a God – that only by having a God is it possible to explain the universe. By offering a plausible competing explanation, they make unbelief a reasonable alternative. To do that, a theory doesn't have to show that it is correct, only that it is as likely to be correct as the 'theory' which says God created the universe. If all explanations for the origin of the universe are equally unfalsifiable, all acts of faith, then one may be as good as another. Physical explanations offer the promise of confirmation by future scientific study and discoveries – all of which sounds more enlightened to late twentieth-century minds than the promise that Christ will return and falsify all competing theories.

Proposals we have been discussing have managed to suggest that we could after all have a universe we can eventually explain and understand all on our own, without need for the idea that there is a creator. If we reduce arguments having to do with 'Is there a God?' to 'Is God necessary?' these proposals give an edge to agnosticism. Where we could have expected to hear the words 'The hand that made this is divine' – the origin of the universe itself – we hear instead 'There doesn't necessarily have to have been any divine hand in this.' Not quite so promising for setting to music.

The question which looms over all this discussion is whether any of these proposals does indeed offer serious competition to the 'theory' that there is a God. Would any of them, if it turned out to be correct, be a complete explanation for the beginning of the universe? Is God, for that matter, a complete explanation?

When a theory requires that we take for granted pre-existing laws or a pre-existing situation or context, and especially when we know what that situation or context would have to be, we haven't really found a complete explanation or a candidate for First Cause which-has-not-itself-been-caused. A pulsing universe needs a previous pulse. And it has to be a universe that obeys a set of laws that cause pulsing to take place. Why should it necessarily be that sort of universe? Where do those laws come from? For 'something' to be more likely than 'nothing' requires a context in which the statistics say this is so. The 'free lunch' requires that the uncertainty principle be in operation. As Polkinghorne asks, 'Who created quantum theory? You don't get it for nothing.'[31] This is no 'free lunch' after all. In all these cases, prior laws or events or boundary conditions – things we don't get 'for nothing' – are necessary. We haven't found a First Cause, and we haven't banished the question of how things just happened to get set up this way. We may succeed in moving the creator back a few steps, but we don't banish the need for a creator – or at least a cause.

Might an underlying system of laws, a situation, or a context itself be the First Cause? Maybe there is something so compelling about the set of laws, the situation, or the context that it brings about its own existence and makes obedience inevitable. If so, by what standard 'compelling'? The answer could be: by the standard of mathematical and logical consistency, only these conditions, laws, or guidelines would satisfy, and these conditions, laws, and guidelines make the universe inevitable. We are far from showing this is the case in any proposal we have reviewed, but let's imagine we could. *Then* perhaps we could say we have reached a First Cause. Mathematical and Logical Consistency dictates that the universe began and developed in this way and no other. Any other way it might have happened – or for it not to have happened at all – is illogical and inconsistent; there is no other choice.

The hope of some who seek a Theory of Everything is just that – that it will be more than a unification of the forces and particles and a set of initial conditions, more even than a unification that will show how forces, particles, and initial conditions are linked. They hope that in the end the only assumption we will need is that there is a fundamental mathematical logic, which could not be otherwise and which makes everything that is real also inevitable.

Could we still insist on asking who invented mathematical and logical consistency? Do we get those for nothing? We could ask that, but we must also remember that we can ask who invented God. At this point it seems we do, indeed, reach a stand-off. If our faith requires that the First Cause be 'scientific' rather than 'religious', it would seem that Mathematical and Logical Consistency is the First Cause candidate of choice.

Could we still believe that God created the universe? Yes, but if God had no choice but to create according to a logic more fundamental than himself, then is God really First Cause of everything there is? We might argue so, if God still had the choice *whether* to create. Perhaps we could have both First Causes simultaneously: God by nature both being and defining 'mathematically and logically consistent'. On the other hand, if God is stronger than any system of logic, if God invents all logic and mathematical consistency, then God is First Cause.

The discussion doesn't end here, but if we confine ourselves to what we've seen so far in this book, we do have a genuine stand-off between two First Cause candidates – God, and Mathematical and Logical Consistency. One can continue speculating, but the bottom line would seem to be that at present we have no scientific way of proving or falsifying either of them, nor are we likely ever to determine the answer by means of the scientific method. To vote for either candidate is a matter of faith.

THE THIRD CANDIDATE

Have Hawking and Hartle now had the temerity to nominate a third candidate . . . the Universe? In their no-boundary model, the Universe just IS, nothing had to create it or cause it. Let us consider the universe as a candidate for First Cause, to find out whether it can join the previous two in this exercise of cosmic one-up-man-ship.[32]

'If the universe has no boundaries but is self-contained . . . then God would not have had any freedom to choose how the universe began', writes Hawking.[33] However, if God didn't have a choice, why is it that Hartle and Hawking DID? Hawking has said that 'the boundary conditions of the universe are that there are no boundaries.' It's true his proposed universe has no boundaries in space or time, but in a sense it still has boundary conditions. One conventional definition of boundary conditions (initial conditions, in this case) is that they are the conditions at the beginning of an experiment – the initial state of everything that's going to be involved in the experiment. But we've also seen other meanings of boundary conditions. Boundary conditions can mean the underlying context of logic and laws, the specification required in order for the proposed situation to exist at all, with no reference to time or a beginning. A universe like the no-boundary universe, without boundaries in time or space but in which neither time nor space is infinite, could in fact exist only if Hartle and Hawking presupposed some rather specific boundary conditions of this second sort.

Hartle and Hawking provided their boundary conditions by giving a specific mathematical formulation which severely restricts the quantum state of the universe – a mathematical formulation which appeals to them on aesthetic and other grounds: it is mathematically elegant, it seems plausible rather than contrived, and it is able to circumvent the need for a singularity. As Hawking has said, this mathematical formulation was not deduced from some other principles of physics. No-one has been able to show that it is the only possible mathematical formulation that is self-consistent and could explain the universe we observe. Hartle and Hawking chose boundary conditions which would apply within their no-boundary universe, which make it possible for such a no-boundary universe to exist. In what way, then, is the no-boundary proposal any different from, or more fundamental than, the other origin-of-the-universe proposals we've discussed? Surely this no-boundary universe also presupposes a context, a situation, a mathematical formulation, without which it couldn't exist. We can still ask: Why this context, this mathematical formulation?

The difference is subtle. Hartle and Hawking's abolition of a 'beginning' becomes a key issue. Their universe needn't be considered as part of a continuum of space, time, or causality which includes anything except itself. If it should turn out that only the mathematical formulation Hartle and Hawking use could have produced the universe as we find it, and if this universe is completely self-contained and self-consistent in both time and space, then the context this universe presupposes is – itself. The question 'Why this context, this mathematical formulation?' can be reasonably answered 'Because this is obviously what IS.' The universe dictates the boundary conditions necessary for its existence – because it exists. What IS, physical reality, becomes a stronger concept than God.

To summarize this complicated argument: Hartle and Hawking are suggesting that we may find that the only way the universe could have got to be the way it is is to have been a universe in which at an instant in imaginary time the time dimension became identical with space dimensions in precisely the way they describe. If this could only happen using the specific mathematical formulation they used, then God didn't have a choice how to create THIS universe, and neither did Hawking and Hartle. If there is additionally no time when the universe didn't exist, then God didn't have a choice of when to create the universe or whether to create it. There are no choices at all.

Before proceeding, we should ask whether a wormhole universe is also a no-boundary universe. It can be considered that, because although there must be a parent universe, the 'time' in which the wormhole forms and the baby universe is born is imaginary time. According to some of its proponents, wormhole theory is a triple threat Theory-of-Everything candidate because it draws together the laws of physics and the initial conditions and even takes a good shot at explaining the constants. We'll see more about that in Chapter 5.

THE MOTHER OF ALL CHICKEN-AND-EGG STORIES

We'll allow those who favour God as the candidate for First Cause to have a first go at knocking this third candidate – the Universe – out of the running.

Don N. Page, a close friend of Hawking's who has collaborated with him on several papers and who lived with the Hawking family in the late 1970s when he was a postdoctoral student at Cambridge, is now a professor at the University of Alberta, Canada. Page is a devout Christian, and he has tried to answer Hawking's question 'What need then for a creator?' According to Page, Hawking has not banished the need for God. In the Judaeo-Christian view, 'God creates and sustains the entire universe rather than just the beginning. Whether or not the Universe has a beginning has no relevance to the question of its creation, just as whether an artist's line has a beginning and an end, or instead forms a circle with no end, has no relevance to the question of its being drawn.'[34] The argument that God not only creates but also continually sustains the entire universe is expressed in the New Testament in a verse in Colossians: 'By him

[Christ, in this case] all things were created . . . He is before all things, and in him all things hold together.'[35]

A circle *does* have a beginning – the moment the artist draws it, or stamps it, or whatever method he or she uses. It is a beginning in the dimension we call 'time', a dimension not contained in the circle itself. For that reason, a circle is probably not a good analogy for the no-boundary universe – a universe which has no time dimension outside itself, such as Page's circle has, in which the 'drawing', or 'stamping', or 'beginning' can occur. Davies indirectly supports Page's view by suggesting that 'Although Hawking's proposal is for a universe without a definite origin in time, it is also true to say in this theory that the universe has not always existed.'[36] Davies is right in the sense that time in the no-boundary universe is not infinite. However, 'always' is a misleading word, for, like many of our words, it has meaning only where there is a time dimension. Hawking insists it's meaningless to talk of a time other than when the universe was in existence. As St Augustine of Hippo said with regard to discussions about a time before time began, 'Let them cease to talk such nonsense!'[37] There was no such 'time'. We might, following Augustine, suggest that God exists outside of both space and time, and *could* create and sustain a universe like the no-boundary universe in which time does exist without a beginning. But it seems the no-boundary universe also *could* exist without there being such a God.

Another question: Even if we find that Hartle and Hawking's scheme is the only way to achieve THIS exact universe – who chose that this exact universe should be the goal? The best rejoinder is that we have this universe and no other, which makes the idea of a choice meaningless. This response wouldn't really counter the proposal that God made the choice in order to have a universe suitable for human beings. In Chapter 5 we'll discuss the anthropic principle, which suggests that even this remarkable suitability is not necessarily a good reason to assume there is a God.

Another idea favouring God as First Cause comes from physicist Karel Kuchar, himself not a believer but a person who obviously enjoys controversy. Not to give himself a choice – perhaps THAT was God's choice.[38] Why should God choose not to give himself a choice, and thus hide the fact that he did? Perhaps

God preferred a universe in which he seems superfluous because such a universe leaves us no gaps, no mysteries where we must assume divine action. Maybe God's choice was to allow us freedom as to whether we will believe in him; God simply doesn't want to be found in the physical universe, because that would intimidate us and abolish our freedom of will. The no-boundary proposal would have been an ever-so-clever way of setting up the universe, if God wished to keep us from discerning his divine hand in creation. But it seems the no-boundary universe *could* exist without there being such a God.

Next, an objection to the Universe as First Cause from those who favour Mathematical and Logical Consistency as the First Cause. The no-boundary proposal presupposes something more fundamental than a particular mathematical formulation. It presupposes that the universe obeys mathematical and logical consistency. Or does it?

There is a way of thinking about it in which the universe-that-just-IS might be a stronger concept than mathematical and logical consistency – might constrain mathematical and logical consistency to be what it is. This is a rather obscure notion which we can best approach by recalling what may be an analogous situation. Space and time were once thought to be absolutes. Then Einstein transformed our thinking about them by showing that massive objects cause a warping of spacetime. As Hawking says, 'Our perception of the nature of time changed from being independent of the universe to being shaped by it.'[39] This statement doesn't sufficiently reflect the fact that this influence is a two-way street. Some lines from one of John Wheeler's poems sum it up: 'Spacetime grips mass, Telling it how to move; And mass grips spacetime, Telling it how to curve.'[40] What is clear is that space and time and the arrangement and movements of objects in the universe can no longer be thought of except as linked. Perhaps that allows us to speculate that, though we may now view mathematical logic as an absolute, we might find that it is not – that it can't be thought of except as linked to this particular physical universe. It could be that mathematical and logical consistency itself is somehow shaped by the way the universe IS.

God just IS. Mathematical and Logical Consistency just IS. The Universe just IS. We might suppose that three First Causes are really one – God, Mathematical and Logical Consistency, and the

141

Universe existing in perfect unity – all defining one another. Short of such an unorthodox trinity, it seems one of the three must be the uncaused First Cause, with no answer to the question why or how. As we end Chapter 4, we can only say that none of the three seems able to knock the others out of the competition. For that matter, have we met the entire slate of candidates?

In Chapter 5 we'll change our approach and try to bring science and religion onto the field in a way which will not allow for a stand-off.

5

THE ELUSIVE MIND OF GOD

In every true searcher of Nature there is a kind of religious reverence; for he finds it impossible to imagine that he is the first to have thought out the exceedingly delicate threads that connect his perceptions. The aspect of knowledge which has not yet been laid bare gives the investigator a feeling akin to that of a child who seeks to grasp the masterly way in which elders manipulate things.

ALBERT EINSTEIN[1]

WE'VE ALLOWED PHYSICS THEORY TO SPIN OUT FIVE TALES OF THE origin of the universe. God hasn't figured in any of them. Instead we've found two candidates to compete with God for First Cause – and no way to cast an objective vote. Since there is probably no person on the face of the earth who *could* make an entirely objective decision among the three – even with much more knowledge than we presently have – without allowing some hidden or not so hidden agenda to weight the decision, let us invite an alien who has never seen our universe to survey this field of candidates. Could the alien, having familiarized himself or herself with our way of doing science and our human assumptions and logic, and then having looked at all the scientific findings and theories and arguments we've seen so far in this book, decide that God is the First Cause of our universe? Surely not. Nowhere in this science have we, with reason's ear, heard the words clearly spoken: 'The hand that made this is divine.' Could the alien decide that Mathematical and Logical Consistency or the Universe is the First Cause? Again, surely not. All three of the candidates are unprovable and unfalsifiable.

The alien might try to help things along by saying: 'The scientific explanations, though short on proof or falsifiability, have made a fair case for thinking that the universe just IS, or that mathematical and logical consistency just IS. Is there a similar case for God? Something more than the negative argument that God can't be disproved?'

That's a fair question. We can go on and on talking about science's inadequacy to assure us a clear, completely objective view of ultimate reality – the chair-as-it-is-in-itself – a vision unbounded by any point of view or context. We can point out that theories like the no-boundary proposal are too speculative to offer a definitive answer, and that even if that weren't a problem, they still beg questions like 'Is this the only origin theory which is mathematically and logically consistent?' We can insist we don't know all there is to know about mathematics and whether it is in itself entirely logical and consistent, and argue that in mathematics there do appear to be logical contradictions. We can cite Gödel and others who show us that truth goes beyond provability. We can even ask, 'But who decided we should have THIS mathematical consistency, this logic, and not another?' These are all reasons why we hesitate to swallow the scientific explanations hook, line, and sinker and conclude there's nothing more to be said. But they're not positive reasons for believing in God.

Have we seen any evidence so far that points to the existence of God? We've seen that in spite of the argument that scientific findings have no inherent moral content, human beings (not least, scientists) have a built-in arrow defining a direction toward truth and away from error, toward knowledge and away from ignorance, toward rationality and away from confusion – attaching a value to these directions. Who or what set this compass? Who or what determined that there be any such orientation at all? We've also seen that the universe on all levels is redolent of design and rationality, and other explanations for this rationality seem slightly contrived in comparison with concluding there has been a Mind at work. These are arguments that there is a God, but they are by no means conclusive arguments. It will take better than that to knock the other candidates out of the running.

Confining ourselves to what we've seen so far in this book, we can say: We haven't found God through this science, but the possibility that there is a God will not go away by means of any vanishing act we've seen science perform so far. Only by an act of faith in God or science could anyone at this point declare which of the First Cause candidates really is the First Cause.

You may be thinking that we have come hardly any farther along on the quest for ultimate truth than we were at the end of Chapter 1. Shall we call it progress that we have cleared the air of any lingering notion that at this level, science can reveal to us the answer to the question: 'Is there a God?' At the beginning of his book *The God Particle,* Nobel physicist Leon Ledermann writes: 'When you read or hear anything about the birth of the universe, someone is making it up.'[2] If Genesis 1 is a work of fiction, so is much of Chapter 8 in *A Brief History of Time.*

Fortunately for our quest, the discussion needn't end here. The moment has come to stop treating religion as a monolith which asks us to accept only that there is a God and that God is the First Cause of the universe. In the pages that follow, we will set other beliefs about God up alongside late twentieth-century science. Can we believe wholeheartedly, without compromise, both in late twentieth-century science and in God – or would such simultaneous belief be double-think or other intellectual dishonesty? *Is* there a conflict, and if there is, *where precisely* is that conflict?

GOD AS THE EMBODIMENT OF THE LAWS OF PHYSICS

When a television interviewer asked Hawking whether he believes in God, he replied that he prefers to 'use the term God as the embodiment of the laws of physics'.[3] American physicist Bryce DeWitt told me 'I think many of us [physicists] see it that way.'

If one interprets God as 'the embodiment of the laws of physics', does that mean one believes 'God' is accessible only to

145

physicists? I've broached that idea informally among some Cambridge physicists and got the resounding reaction that that would be 'a rather poor sort of God'. Perhaps they were only pretending to be modest. However, I haven't found that most physicists really fancy themselves as high priests, no matter how much some of the rest of us insist on regarding them as that.

Since Hawking doesn't choose his words casually, it seems not too pedantic to ask what the word 'embodiment' implies here. Hawking surely isn't thinking of an actual incarnation of the laws of physics in a Person. His statement could mean that when most people speak of God, they are actually only anthropomorphizing the laws of physics. Or he could be suggesting that the laws have a life or creative force of their own – as we speculated in Chapter 4 when we said that the equations might BE the 'fire'. 'God as the embodiment of the laws of physics' undoubtedly means different things to different people. It probably most often reflects the following point of view: all I can know about the power that drives the universe is what I find in the laws of physics, so I must leave it at that; for all practical purposes besides pointless speculation, 'God' is the name we have given the laws and the pattern.

There is obviously no contradiction between this idea of 'God as the embodiment of the laws of physics' and a belief in science. However . . .

A PRESENCE BEHIND THE PROCESS

Suppose one believes in God not as the embodiment of the laws of physics, but as the *source* of them, a God behind and beyond the laws – or, even more fundamental than that, the creator of a context in which such laws would inevitably arise and make a universe. This God needn't be a person. A Mind perhaps, but we shouldn't expect to have a word or concept that fits. Einstein wrote of his profound reverence for 'the rationality made manifest in existence' and also of attaining 'that humble attitude of mind towards the grandeur of reason incarnate in existence, and which, in its profoundest depths, is inaccessible to man.' He goes on to say that 'this attitude appears to me to be religious, in the highest sense of the word.'[4]

This view reverses Hawking's order: God is no longer the embodiment of the laws of physics. The laws of physics are the embodiment of a more fundamental 'rationality' – to which we could give the name 'God'. 'Reason', 'in its profoundest depths', is beyond our reach – unimaginable and unattainable in a way the laws of physics are not. We can imagine such a God existing if the universe and the laws of physics didn't exist at all. Hawking's God could not. But Hawking seems to hover on the verge of Einstein's philosophy when he asks 'What is it that breathes fire into the equations and makes a universe for them to describe?'[5] and when he suggests that *after* we understand the Theory of Everything, then we can go beyond that to the 'ultimate triumph of human reason', knowing the Mind of God. This last triumph is the one Einstein thought unattainable.

Belief in this God isn't necessarily a belief that God had any particular purpose in creation or continues to be involved in it in any way. All it suggests is that there is (or once was) what Einstein calls a 'rationality' – what Hawking calls a 'Mind of God' – and that its existence somehow resulted in the universe – full stop.

Does science contradict belief in this God? No. Any theory, no matter how fundamental its scope, which begs such questions as 'Why should nothingness be unstable? . . . Why should uncaused events be possible rather than impossible? . . . Why this mathematical consistency and not another? . . . Why these laws and not others? . . . Doesn't it still look as though a Mind had to make a choice?' can be used to invoke this God. So can a reverence for the rationality of the universe. This is the God who competes with Mathematical and Logical Consistency and the Universe at the end of Chapter 4. Not proved by science, but also not ruled out.

Belief in this God answers the question 'Who-done-it?' but not the question 'Why?' What if God had a motive?

THE LEAP TO PURPOSE: THE GOD WHO WISHES TO DRINK TEA

John Polkinghorne provides us with the following analogy.[6] Suppose you happen to find a kettle of water boiling on the stove.

You might ask, 'Why is this water boiling?' And someone might answer, 'This water is boiling because the combustion of hydrocarbons has generated heat which has heated up the water until the vapour pressure is the same as the atmospheric pressure and so the kettle boils.' That answer would be correct. Your next question would probably not be 'But who invented those laws?' – unless you have been reading this book too long and really need a cup of tea! If you did ask that question, that would be invoking the God we spoke about above, the presence behind the process. The answer you're more likely looking for is something simpler, such as 'This kettle of water is boiling because old Granny intends to have tea' – the answer that supplies not only 'How' and 'Who' but also 'Why'.

Why is the kettle of the universe boiling? Does Someone want to have tea? Is there a motive?

It is intriguing to speculate about that question, and many have done so – some less seriously than others. Perhaps the universe is someone's scientific experiment. Perhaps it *was* an interesting experiment, abandoned now to boil itself to nothing. Perhaps it's a labour of love . . . or whimsy. Perhaps it came about because of an artist's compulsion to create. Perhaps it's the work of a master engineer, or an inventor who delights in setting up elaborate clockwork. Perhaps someone was lonely. Or just bored with eternity on his hands. You've probably heard: 'God is not dead, he's just off working on a less ambitious project.'

Does attributing a motive to God contradict evidence we find in science?

With the tea-making analogy we raise again matters we talked about at the end of Chapter 3. Science might explain the physical process involved in the creation of the universe (like the physical description of how water boils). But the universe can't be shown to be *only* the sum of the physical processes involved, no matter how well these processes explain all physical phenomena. Even among physical explanations, we can never know we have found the simplest, the most fundamental, or the most significant explanation. We can only find out we *haven't* – when and if we discover an explanation that is simpler, more fundamental, or more significant.

It doesn't require double-think to accept both explanations for a boiling kettle at the same time – the physical explanation and

148

the explanation in terms of purpose. We aren't forced to choose between them. It also doesn't require double-think to believe that there is a complete physical explanation of the universe, self-consistent and self-explanatory – and at the same time to harbour strong suspicions that this explanation is not the most significant explanation. Science does not rule out such a possibility. The fact that we don't need another explanation doesn't guarantee there isn't one.

When we become more specific about the 'motive' – and particularly when we insist on making ourselves part of the purpose – we find ourselves in deeper water.

THE WATCHMAKER

The popular picture of the evolutionist vs. God-as-creator debate – which sees it as centring on whether or not Genesis is a literal account of creation – obscures a much more basic issue. Even liberal and intellectual-mainstream Christian and Jewish beliefs, as well as beliefs of other world religions, appear to collide with science over the question of whether beings such as ourselves, apparently capable of responding to God, could be a part of God's purpose. Is evolution a killer of an obstacle for belief that God could have had the intention of creating *us*?

It isn't possible in a few pages to do full justice to a subject which has been the focus of so much thought and controversy for nearly a century and a half. Volumes have been written about evolution and its implications. The most successful popular book about evolution – Oxford biologist Richard Dawkins' *The Blind Watchmaker* – took 318 pages and a computer program to get his argument across. Dawkins explicitly states that he intends the book as a challenge to belief in a designer God, and he carries that further to challenge belief in any God at all. It is a brilliant book and a brilliant challenge. In the spirit of this chapter, we will not ask whether Dawkins' science is correct. (It is orthodox science.) And we will not ask whether belief that God has or had a purpose is correct. Our question is whether it is possible to accept both explanations without double-think or hypocrisy, and if not, why not. Dawkins provides us with a superb statement

149

with which to 'test' religious belief. If his interpretation of evolution is not destructive to belief, it is difficult to see that any interpretation could be!

The specific argument for God which Dawkins sets out to undermine is the 'argument from design'. It was eloquently expressed by William Paley, a theologian and naturalist of the eighteenth and early nineteenth centuries who was also trained as a mathematician. Dawkins tells us that he shares Paley's reverence for 'the complexity of the living world . . . [Paley] saw that it demands a very special kind of explanation.'[7] Paley wrote that if you or I discovered a watch lying on a heath, didn't know how it had come to exist, and studied it closely, we couldn't help but conclude that 'the watch had a maker: that there must have existed, at some time, and at some place or other, an artificer or artificers, who formed it for the purpose which we find it actually to answer; who comprehended its construction, and designed its use.' Paley went on to say that 'every indication of contrivance, every manifestation of design, which existed in the watch, exists in the works of nature; with the difference, on the side of nature, of being greater or more, and that in a degree which exceeds all computation.'[8]

So far, Dawkins and Paley are completely in accord. But Paley insisted that all this evidence pointed unmistakably to the existence of a Creator God. According to the argument from design, not only the astounding complexity of nature but also the miraculous way the environment provides for all creatures, including humans, is evidence of an all-powerful and caring God. 'We've come a long way since Paley's watch,' writes Paul Davies on a condescending note in *The Mind of God*.[9] Have we indeed? Dawkins insists we should all hope that we haven't lost our sense of wonder at all the evidence of complicated design we find around us in nature. However, Dawkins also insists that Paley's explanation for that evidence was wrong: Darwin discovered a more powerful and convincing explanation.

Dawkins is somewhat misleading in speaking of the 'argument from design' as 'always the most influential of the arguments for the existence of a God'.[10] Few modern believers in God place the argument from design high on their list of reasons for belief. Arguably few believers ever have. It would be more accurate to say instead that the argument from design *was,* for a brief period

150

in the eighteenth and nineteenth centuries, a popular argument for *intellectual* believers to use in debates and discussions with *intellectual* unbelievers, because it seemed to offer independent and scientifically acceptable underpinning for belief. But already in that pre-Darwinian world there were religious writers such as John Henry Newman who pointed out that the argument from design was convincing only to those with a pre-existing faith and might also be used effectively as an argument for atheism. Newman thought that even using it as secondary evidence tended to give it too much importance and detract from the *real* foundations of faith.

However, it would be incorrect to conclude that Dawkins merely does battle with an outdated straw man. In fact, he launches a much more sophisticated and threatening assault on religious belief. With the theory of evolution on the one hand and the argument from design on the other, we have two explanations for the complexity of nature – including human beings. We know that no explanation can make a legitimate claim to be the only possible explanation. We have seen that sometimes the only reasonable solution is to allow a temporary stand-off between competing explanations, while not assuming that all possible explanations have even been suggested yet. We have also seen that it is often possible to accept two explanations at once. It didn't require intellectual dishonesty to accept the two explanations for the kettle of boiling water. Perhaps one might say, then, 'What does it matter whether God created human beings in the way described by Genesis or in the way described by Darwin? The important thing is that God created us.' What would be the problem with believing that evolution is a process which God invented, has used, and continues to use to achieve his ends?

While it is often possible to accept two alternative explanations and even combine them, not all pairings of explanations are allowed. We might find in some instances that one of the explanations, with a wealth of evidence to support it (as the physical explanation for the tea kettle has), *rules out* the alternative explanation. Suppose the physical description of how water boils simply did not allow Granny to put the kettle on the cooker? This is Dawkins' challenge. He insists that evolution rules out the religious explanation. What sense is there in *any* argument for belief in a God who set out with the purpose of 'creating Man', if evolution isn't merely an alternative explanation but has actually shown that God *could not* have 'created Man'? Dawkins' argument is that the theory and the evidence of evolution show us that the universe is a universe which could not possibly have a designer. Natural selection is a process that rules out the possibility of having a goal or a pre-specified end-product such as the human eye, or creatures like ourselves that can ask questions like 'Is there a God?'

Yet the evolutionary process is not entirely random. If it were, the complexity we find in such organs as the human eye (or even in much less complex items) and the superb way environments are suited to support their inhabitants would beg a designer. As Dawkins puts it, 'Living organisms are well fitted to survive and reproduce in their environments, in ways too numerous and statistically improbable to have come about in a single chance blow.'[11] We hear tell of getting enough monkeys at typewriters and giving them enough time and having one of them type the works of Shakespeare. We speak of 'enough monkeys' and 'enough time'. But we'd actually need far more monkeys than have ever existed in the universe – and far more than the 10 to 20 billion years since the universe began – to give us anything like good odds of this happening. In spite of all the literary squabbling about whether Shakespeare really wrote the works of Shakespeare, odds on it *was* Shakespeare, not monkeys.

There are many living things and parts of living things that are more complex than the complete works of Shakespeare, and the idea that these things sprang into being by pure chance is not acceptable. However, the odds can be improved considerably without introducing God.

Dawkins tells us: 'The Darwinian explanation, of course, involves chance too, in the form of mutation. But the chance is filtered cumulatively by selection, step by step, over many generations.'[12] In order to clarify what he means, and why the 'step by step' is all-important, Dawkins begins not by making the complete works of Shakespeare the goal, but just one phrase from *Hamlet* – 'Methinks it is like a weasel'. That in itself is too much to expect to get easily from the randomly-typing monkeys, though the odds are better than for the complete works. Dawkins doesn't have a monkey and he doesn't think his readers are likely to either, so he programs his computer to generate a random phrase of 28 letters and spaces, the number of letters and spaces in the goal phrase. If he wanted to go on simulating the monkey situation, he would allow the computer to continue generating 28-letter sequences at random, hoping, against stupendous odds, that one of the sequences would read 'Methinks it is like a weasel'. Instead, he sets about short-circuiting the odds. It may interest you to know that computer programs similar to the one we are about to discuss are being used experimentally to make business decisions and pick stocks.

Setting up a computer program and a goal phrase, 'Methinks it is like a weasel', might be misleading, because it implies that someone did set up a super-complex computer program to achieve our present-day universe and us. We have to play God to see how this works, but, as we shall see, that doesn't necessarily mean that the process requires a God to set it up or allows God to do so. Dawkins wants to show us that nobody had to plan, predict, or foretell such things as the human eye or the flower seed or human beings; indeed, that nobody *could* have done so.

Twenty-eight randomly generated letters and spaces on Dawkins' computer came out

WDLMNLT DTJBKWIRZREZLMQCO P

Dawkins programmed the computer so that it would repeat this same 'parent' sequence of 28 characters over and over, and very occasionally (at random) allow itself to make a mistake in copying. The result was a list of many sequences that were replicas of the parent sequence, with here and there among them a letter sequence which had a mistake in it – making this 'offspring' slightly different from its brothers and sisters. We are asked to think of the mistakes as corresponding to mutations in the real world. A 'mistake' doesn't always make for inferiority.

Dawkins' computer studied the letter sequences, including those with mistakes, and judged which among them most resembled the goal phrase. This is not a moment for over-optimism. None of them in the least resembled the goal phrase, a fact which is extremely important to our understanding of the process of evolution. We have to look closely even to see any difference from the parent phrase. But one of the mistakes does happen to be just a hair's breadth closer to the goal phrase than its brothers and sisters. The winner was:

Dawkins carried out this procedure repeatedly, a procedure which we can sum up as follows. The computer behaves like a copying machine and produces many copies of the 28-letter parent sequence, making some random mistakes in a few of the copies. The computer studies all the letter sequences of this new generation and chooses from among them the one sequence that most resembles, by even the smallest margin, the goal phrase. The computer uses that sequence as the new parent sequence and begins the process over again by making many copies of it. After ten 'generations' the sequence that most closely resembled the goal phrase was

MDLDMNLS ITJISWHRZREZ MECS P

After 20 generations it was:

MELDINLS IT ISWPRKE Z WECSEL

which looks a little more promising! After 30 generations it's

METHINGS IT ISWLIKE B WECSEL

And 40 generations gets us almost there:

METHINKS IT IS LIKE I WEASEL

Generation 43:

METHINKS IT IS LIKE A WEASEL

Dawkins tells us that the process took his computer (using a rather slow, old-fashioned master program) less than an hour – a vast improvement over the monkeys or having the computer merely continue to generate random 28-letter sequences.[13] The odds for achieving complexity and something that looks as though it had a designer have improved considerably.

But in this exercise we *did* have a designer with a design ('Methinks it is like a weasel') in mind. The computer was programmed to choose the winner on the basis of resemblance to that goal design. Our question is did anybody set up a program like that in the real world to choose the winner on the basis of resemblance to a flower seed, or a resemblance to you or me? Could anyone have set up a program like that in the real world? Did God do that?

Most of us learned in biology class that in the real world the 'winner' in each generation is the offspring which, because of the very slight difference in its make-up (and we are talking VERY slight difference), succeeds just the tiniest bit better in surviving and producing offspring in the environment where this generation finds itself. It doesn't have to be able to do a great deal better, just the most microscopic hair's breadth better.

In order to arrange things in the real universe to reach a predetermined goal (in the way Dawkins used his computer), one would have to pre-arrange the environment in which the competitors for 'winner' are attempting to survive and reproduce. The exact conditions of the environment are all-important in determining which competitor has the edge and ends up 'winner' – parenting the next generation. What would end up 'winner' in one environment would not end up 'winner' in a different environment. Could anyone (God) manipulate the environment like that? It would be a little like fixing the outcome of a beauty contest by painting a portrait ahead of time to show what the judges would consider beautiful. But 'fixing' the environment would be far more difficult. It would also be far more difficult than setting up a computer program to determine what sort of competitor would win.

What creates the environment? The environment is not something that sits waiting to have mutations tested against it. It is the product of myriad other lines of evolution that have been working by exactly the same process – all testing their mutations against one another in an extremely complex and always changing network. The magnitude of the problems involved with trying to set up the environment in a particular way to favour one sort of mutation is almost beyond comprehension. The history of any environment has had as part of it the random 'mistakes' in the copying, and the environment has not controlled which mistakes would occur. Mutations are not called up by the environment to suit its needs, or by the need of an organism. An eye might be badly needed and give an organism a tremendous advantage, but if the particular 'mistake' that leads to an eye never occurs, there will be no eye. With the static, artificial computer 'environment', designed as it was to make its choices with the goal phrase in mind, it was possible to start out with the goal of 'Methinks it is like a weasel' and arrive there. Without that artificial environment, no-one could arrive at a pre-arranged goal.

However, evolution does produce recognizable design, because it produces, by the process of elimination, beings that can successfully survive and produce offspring in the environment in which they find themselves. Why does an organ like the eye seem so superbly suited to answer our human needs within our environment? Because it was such needs, changing and shifting over time along with the ancestors of the eye, which fine-tuned every choice of 'winner'. It's bound to look as though it were designed as a masterpiece of equipment for this environment, or that the environment was designed as just the sort of environment where the human eye would work well and be very useful. Because there is often competition of many sorts for survival – and because there is an extremely long time for all of this to occur – we find finely tuned design and complexity of a very high order indeed among a myriad of species. The jump directly from primordial soup to the human body is unimaginable except as a miracle. Taken step by step it is a plausible natural process, though no less amazing in what it accomplishes.

Some questions spring to mind. For instance, isn't the real universe, like the computer, a controlled world – controlled by the laws of physics?

The constraints of gravity, the existence of light, these haven't evolved, at least not on earth since this earth began. If the laws of physics manifest themselves differently on other planets, as we know they must, then evolution on those planets will have taken a different course, perhaps dramatically dissimilar to the course it has taken on earth. Obviously the physical laws do powerfully influence and limit the course of evolution. But an environment consists of much more than the laws of physics in their more obvious manifestations. There are also many different ways an organism can find to cope with the constraints of the physical laws. We should not overestimate the influence of these laws in determining how a species evolves. As we shall learn when we return to this matter in Chapter 6, the laws leave immense freedom.

A second question: Has there been enough time for this process to produce something as complex as the human eye? Since there probably hasn't been a significant change in the eye since long before recorded human history began, this seems a legitimate question. However, there has been far more time than it's possible for us humans to think about in any way that makes sense to our intuitive feeling for time. Biologists are able to demonstrate that there has been ample time even given the extreme slowness of the step-by-step process.

Regarding all the minute alterations that were required along the way in order to arrive at something as complex as the eye: it's difficult to see how some of those changes could possibly by themselves have provided an edge for survival. But biologists tell us we grossly underestimate how much of an advantage can be provided by a microscopically tiny change.

The theory of evolution is a well-established theory, not speculative and on the fringe of science. The fossil record provides supporting evidence. It is true that fossil evidence, like astronomical observations, cannot be controlled in the way a laboratory experiment can. We have to take what's on offer and make the best of it. But gaps in the fossil record also provide interesting and telling evidence of the way the evolutionary process works. Easier to control are studies of still living species, many of them microscopic, whose generations pass much more rapidly than human generations do. The remaining problems with the theory are mainly a matter of working out details and deciding in certain areas which interpretation of the evidence is more likely to be correct. The consensus of science today is that the theory of evolution is a powerful theory indeed, extremely well supported by evidence.

We have yet to ask the question: Where did the process of evolution come from? Did it require an inventor? Darwin didn't attempt to answer that question, nor did he attempt to account for the first few living forms. We would be justified in insisting that the process of evolution and the marvel of DNA are at least as astoundingly clever and impressive as any of the complexity and design they have produced. If anything in nature compels us to ask with awe 'WHO thought THAT up?' it is this process.

With the computer model, we didn't ask who the programmer was, or why the program made all those duplicate copies, or why it made a few random mistakes, or why there should be any program at all. In the real world, we are entitled to ask such questions. Dawkins built into the computer program the basic ingredients of the process. The properties of DNA turn out to be the basic ingredients needed for this process to work in the real world. These basic ingredients, probably originally in a much more rudimentary form than we find in DNA, had to come from somewhere, or else life and the selection process couldn't have got started.

157

There are three basic ingredients. The most basic is the ability of something to make copies of itself. The 'somethings' with this ability are called replicators. The second ingredient is that the replicators must make occasional mistakes in the copying. We'd think of that as a disadvantage in normal circumstances, such as when we photocopy a letter, but we've already seen that in evolution it's essential to the process. Whether this ingredient will normally arise from the first ingredient isn't certain. The third ingredient is that there must be something about the replicators that has an influence over how likely they are to be replicated.

How did these ingredients come into existence in the first place? There are several theories explaining how they might have arisen in the universe. In some of the theories it is statistically more probable that they will arise than it is in others, and there is disagreement among experts as to which set of statistics we ought to accept. Dawkins asserts that even at their worst the statistical odds of the ingredients and the process arising spontaneously in the universe are good enough to explain their arising, without having to posit an inventor or an instigator.

If Dawkins is right, then the absolute minimum requirement for life to arise in the universe is that we have a universe in which life *could* exist, a universe in which the statistics are as favourable as they are, and a universe in which we have the laws of statistics as we know them.

Where do these minimum requirements come from? A little later in this chapter we shall be studying the odds of having a universe where life could exist and where the statistics are as favourable as they are. Asking where the laws of statistics come from, we find our way back to one of our previous First Cause contenders – mathematical and logical consistency. We reach our previous stand-off.

Dawkins has this to say about such stand-offs: 'To explain the origin of the DNA/protein machine by invoking a supernatural Designer is to explain precisely nothing, for it leaves unexplained the origin of the Designer. You have to say something like "God was always there," and if you allow yourself that kind of lazy way out, you might as well just say "DNA was always there," or "Life was always there," and be done with it."[14] That sounds rather familiar! In principle we could not disagree with Dawkins. Mathematical and logical consistency was always there . . . the

158

universe was always there (in the sense that there was no time before the universe) . . . God was always there. However, though it may be the lazy way out to say we don't know which it is on the level of DNA and Life, when it comes to First Causes, there is no other objective way to proceed.

Dawkins, however, has an interesting additional reason for arguing that the answer is not God. It rests on an assumption that we saw in Chapter 3, that simplicity is at the heart of everything. If we have a relatively complex candidate for First Cause, then by this criterion it must give way to a simpler candidate for First Cause. Dawkins writes, 'Any God capable of intelligently designing something as complex as the DNA/protein-replicating machine must have been at least as complex and organized as that machine itself. Far more so if we suppose him additionally capable of such advanced functions as listening to prayers and forgiving sins.'[15] Bernstein's declaration that 'God is the simplest of all' seems to fall apart here. We have to admit that if God did design all this, God is nobody's simpleton. If simplicity is the clue for finding out what is First Cause, mathematical and logical consistency might seem to win hands down over Hawking's universe that just IS or a God that just IS. However, we must caution ourselves that the notion that God would have to be 'complex' to perform all these functions is an anthropomorphic notion, as much so as the notion of a God who can't notice anything so small and insignificant as human beings.

We must now at last get to the bottom line of this section and examine the question of whether, given the way evolution works, it would be reasonable to believe that God created us. Even if we believe in God as the First Cause of the universe – if God, not mathematical and logical consistency, instigated the situation in which the process of evolution would arise – can we possibly allow ourselves to believe that this God coaxed us out of a process which seems so ingeniously designed to produce an undesigned universe?

At the risk of appearing sacrilegious, let us put you in the position of God, and ask the question: How will you, God, use your wonderful invention – evolution – to produce a being like me? We'll suppose that you have the universe in operation already, with the statistical odds weighted in favour of life arising. The DNA-replicating machinery has got itself going. That seems a lot to assume, but it appears we can assume it regardless of which First Cause we choose, and we are choosing just now to assume it will work for you, God. We have the first flicker of life in the primordial soup. How will you start with that and end with this woman and others like her who will be so confoundedly curious about whether you exist?

Perhaps for a while you, as God, might allow things to drift along to see what interesting creatures this process of evolution will produce. The number of paths it might follow are infinite. Will it produce every possible creature? If so, you can relax in the certainty that we humans will appear eventually. But the living creatures we have in the world do not represent every possible form of living creature that could exist here, given the laws of nature. It's safe to assume that there are many theoretical animals that evolution could have produced – that could be surviving and competing quite successfully in this world – but that it hasn't produced. Why one then, and not the other?

Let us return to the sequences of letters in the computer simulation. In each generation the computer randomly produced some errors, but not every possible error that it might have produced. In the real world, mutations appear at random, but not every possible mutation. Far from it. Remember that the mutations *are* random, not a response to the needs of the environment or the need of the creature to survive. Many mutations won't be advantageous.

The role of the environment is to determine the winner whose winning advantage will be passed to the next generation. The field of candidates is limited to the mutations provided by random chance, and there is no guarantee that any of the candidates will survive. 'The judges reserve the right not to award the prize if no candidate is deemed suitable.' There's certainly no guarantee that the best of all possible mutations (for this environment) will appear as a candidate. If three eyes would be better than two but the 'mistake in the duplication' that might later lead to the extra eye never turns up in any mutation, there's nothing for it but to make do with two eyes.

Returning to your role as God. Will you intervene by secretly causing certain mutations to appear? You could do that and nobody would know the difference. To the biologist the mutations would still appear to be random. It's a puzzle why you'd invent this marvellous process and then circumvent it, but you're God, and there might be a good reason that we humans haven't thought of. We already saw that it would be very complicated to manipulate the environment which chooses the winner in each generation. But again, you're God, so you must surely be capable of thinking of a way. Perhaps you even created a world rife with false evidence for evolution. But that also leaves us with troubling questions about you and your devious motives. Would you have to do any of that to create beings who will ask such questions as 'Is there a God?' and who will respond to God? No. Not if Richard Dawkins and orthodox science are correct.

We must assume for the sake of this argument that it will not matter to God whether I and others like me, who have the capacity to ask whether there is a God and to respond to God, have two eyes or three or whether we fly, swim, walk, or perhaps slime along on our bellies. Why should that matter? Let us ignore the silly quibble that sliming along on our bellies would not be creation in God's image. 'Image' certainly involves more than physical appearance and mode of transportation. I believe I could still be everything that I think of as essentially 'me' and slime along on my belly, especially if all the other attractive creatures around me were doing the same.

Asking such questions as 'Is there a God?' and responding to God would seem to require self-awareness and a certain level of intelligence. It almost certainly does not require the specific species we know of as homo sapiens. It requires a certain level of evolutionary development, not a specific evolutionary path or a specific evolutionary goal.

Dawkins and most of his colleagues conclude that the emergence of self-awareness and intelligent beings is extremely probable. Some critics call this only a pet assumption, left over from the nineteenth century, that evolutionary 'progress' is inevitable. However, though evolution doesn't always move toward greater complexity, that is the overall trend. It seems clear that in a universe like the one in which we find ourselves the evolutionary process described by Dawkins would with enough time inevitably produce creatures that were self-aware and of great enough intelligence to ask questions such as 'Is there a God?' We know that such a creature has arisen once on our planet.

Penrose, the Oxford physicist and mathematician whose work

on black holes and singularities we discussed in Chapter 4, has this to say in *The Emperor's New Mind:*

> It seems to me to be clear that the musings and mutterings that we indulge in, when we (perhaps temporarily) become philosophers, are not things that are *in themselves* selected for, but are the necessary 'baggage' (from the point of view of natural selection) that must be carried by beings who indeed are conscious, and whose consciousness has been selected by natural selection, but for some quite different and presumably very powerful reason.[16]

In answer then to the question whether God could let evolution take its course (without any secret manipulation) and be fairly certain – maybe absolutely certain – of ending up with creatures who would be advanced enough to ask whether there is a God and respond to God: orthodox science and the current mainstream interpretation of the theory of evolution declare the answer YES. Whether or not God did any such thing, and whether we need a God to explain the emergence of such creatures, are not questions for this chapter. We can only conclude that if our knowledge of the observable universe is reasonably correct, if there is a God, and if evolution works the way most evolutionists, including Dawkins, think it does, not only would God have been able to 'create Man' through evolution, God would have had to do some fancy manipulation behind the scenes in order to change his mind and PREVENT such a creature from emerging.

Experts in other branches of science, particularly the areas of chaos and complexity which we examine more closely in the next chapter, are not entirely ready to agree that the emergence of life is so probable. When it comes to the question of how such levels of organization as DNA arise in the universe, our understanding is in its infancy, and many feel it is much too soon to say how probable such a development is. By some calculations, life as we know it turns out to be not so probable as Dawkins believes. Not so inevitable, but a little contrived. Barrow speculates that there may be a still undiscovered organizing principle, in addition to the laws we know, which dictates the evolution of complex systems. Whether there is, and what it is, are at present a great mystery. It's tempting to speculate that this organizing principle might be God, but it could also be something else science will discover and explain.

We are left with a question-mark regarding how likely or unlikely it is that the universe should exist in the way we observe that it does, a way which allows for the emergence of sentient beings and even makes their emergence probable. If it is likely, then we might conclude that God, having set up this likelihood from the beginning, has watched it all unfold with no further tampering on his part. If it is *not* likely, then it's difficult to refute the argument that we need God to explain why we should have a universe so against the odds. Either way, God, though perhaps not ruled *in*, is certainly not ruled out.

Paley didn't point only to complexity such as the human eye as evidence that there is a God who designed the universe. He was a mathematician as well as a naturalist, and he regarded the laws of gravitation, without which life as we know it could not exist, as another strong piece of evidence. That second sort of 'argument from design' is not dead. It is alive and well and living in physics.

THE UNIVERSE AS A 'PUT-UP JOB'

Before the mid-sixteenth century, both religion and science taught that our earth was the centre of the universe. In 1543 Nicolas Copernicus published his sun-centred astronomy, showing that a far simpler and more convincing explanation for all the patterns of movement we observe in the heavens is that the earth is a planet orbiting around the sun. We now know that the sun, in turn, orbits around the centre of a galaxy. We also think that from any vantage point the universe looks more or less the same, on the large scale, as it looks from earth, and no vantage point has any more or less right than any other to think of itself as the centre. The study of evolution tells us that the natural environment on earth was not necessarily designed with us humans in mind. All of which certainly takes us down a notch!

Nevertheless, the more we discover about both the cosmic and the microscopic levels of the universe, the more we seem to find ourselves again mysteriously reinstated as the kingpin. Laws and constants had to be set up with incredible precision at the instant of creation, or we couldn't be here. These statistics don't appear to work at all in our favour. We can't escape the impression that some careful planning and exquisitely intricate fine-tuning must have occurred with us specifically in mind. Is the universe, as British astronomer Fred Hoyle dubbed it, a 'put-up job'?

For example: If the electric charge of the electron were only slightly different, stars wouldn't explode in supernovas to fling back into space the raw material for new stars like our sun or planets like earth. If gravity were only slightly less powerful than it is, matter couldn't have congealed into stars and galaxies, nor could galaxies and solar systems have formed had gravity not been at the same time the *weakest* of the four forces. If the expansive energy (resulting from the Big Bang) and the force of gravity had differed from equality by more than 1 in 10^{60} at a time less than 10^{-43} seconds after the Big Bang (about the earliest moment at which time and space have any meaning), the universe either long ago would have collapsed again to a Big Crunch, or else there would have been such run-away inflation that gravity wouldn't have been able to pull any matter together to form stars.

The electric charge and the mass of the electron, the strength of the gravitational force, in fact of all four of the forces, and the values of other constants, are values which no current theory can predict. They are discoverable only through observation. And yet these values and the relationships among them have to be precisely right – to an incredible degree of precision – to allow for our existence. Even if the emergence of life in this universe is by some estimates highly probable, having the particular universe in which such a probability exists is highly *im*probable.

Is the universe a great conspiracy to make intelligent life possible? There is nothing coming from science to rule out this explanation. Once again, however, there are possible alternatives.

SECOND GORDIAN KNOT: THE ANTHROPIC PRINCIPLE

The 'anthropic principle' is probably the strongest argument ever made that we can never observe the universe other than from a biased point of view – and that our point of view dictates what we find. The principle suggests that we find the universe to be as it is because we exist, which is a slightly backwards way of saying that if the universe were different we wouldn't be here to notice it. We see such values as the charge of the electron, the strength of the gravitational force, and the cosmological constant as they are because if the values were different we wouldn't be around to observe anything. There seems to be no way we *could* possibly observe a set of values that was very different from what we do observe. Having said that, it's difficult to see we've said anything at all. Some scientists and philosophers insist we haven't.

The anthropic principle has several forms. In its 'weak' form, it can be used to explain why conditions are just right for us to be here on the earth at this particular time in the history of the universe. The answer isn't difficult, given the fact that we are here (which we're allowed as a given). If the conditions weren't just right, here and now, then we wouldn't find ourselves here now, but somewhere else, at some other appropriate time – asking ourselves the same question. There would be intelligent life around only at the specific time and place that allowed for it. That's not too hard to swallow, as long as we don't go on to ask why conditions in *any* time or place should be just right for us.

The 'strong' anthropic principle tries to answer that question. We can picture a very large set of different, separate, possible universes. The conditions in most of these possible universes would not allow the development of intelligent life. In a very few of them – perhaps only one – the conditions are just right for stars and galaxies and solar systems to form, and for intelligent beings to develop and study the universe and ask the question 'Why is the universe as we observe it?' The only answer is 'Because if it weren't, we wouldn't be around to ask the question.' However many, or however few, possible universes there are around, there *must* be at least one which allows life to emerge in it at some stage of its history.

Carry this a step further and you put human beings in a highly enviable position. The fact is that according to some current calculations the odds against having a universe at all are so astronomical as to make the odds of having a universe 'like ours' (once we've got a universe) look like a pretty good gamble. Ask the question again in a slightly different manner: 'Why do we observe a universe at all?' And again the only answer might be 'Because if we didn't we couldn't be around to ask the question.' In other words, there is a universe because we exist. That's close to what Wheeler meant when he suggested that there can be no physical laws at all unless there are observers to figure them out. We could get a very big head about this. We could even start assuming that WE are the First Cause. There is a counter-argument that the odds of having a universe at all are *so* poor that you shouldn't really even use the anthropic principle to explain the existence of the universe.

Interestingly enough, though the anthropic principle competes with the interpretation that says there must be a God with us in mind, and seems to erase the 'need' for such a God, it isn't from the religious community that the strongest resistance to the anthropic principle comes. It's from physicists.

HACKING AT THE SECOND GORDIAN KNOT

Must we resort to the anthropic principle to try to explain the universe and what seems to be such hair-splitting fine-tuning in its initial state? Is this the only possible answer other than 'God did it?'

Let's begin by finding out why so many scientists dislike the anthropic principle as an answer. First of all, it's the kind of thinking that sets us going in circles and arriving nowhere. 'We are here because the universe is as it is and the universe is as it is because we couldn't be here if it wasn't and obviously we are here' doesn't tell us very much more than saying 'Well, here we are' and leaving it at that. If that's the sort of ultimate answer we're going to discover, our quest is hardly worth the trouble.

Second, the anthropic principle is another door slammed in our faces. As Hawking puts it, 'Was it all just a lucky chance? That would seem a counsel of despair, a negation of all our hopes of understanding the underlying order of the universe.'[17] If we were doing a scientific experiment and produced a result this much against the odds, we certainly wouldn't dismiss it as an accident or a bit of luck. We would be obliged to look for a cause.

Several theories have suggestions to offer:

166

THE INFLATIONARY UNIVERSE

When my brother and sister and I were small, my father used to astound us with a game he called 'Mystery Number'. He would ask us to confer among ourselves, agree on a number, and not tell him what it was. Then he would give instructions for a series of mathematical operations. For instance he might say 'Add two', wait while our little mental computers clicked, then say 'Subtract one', then 'Multiply by four' and so forth. We were to work all these out in our heads without sharing our answers with him. We noticed there was no particular sequence which he followed. He called out these orders randomly. This would continue for a little while and then he would announce the result we'd arrived at. Amazement! How could he know the answer at the end of our train of calculations if he didn't know what number had started it off? None of us thought to ask 'But do you know what number we started with?' We assumed he must.

One day he made the mistake – or perhaps it wasn't a mistake but his way of teaching us – of having us not agree on a number but think of separate secret numbers. At first it seemed all the more mystifying that he could know the answer at the end of the chain of calculations, when we had each started with a different number, but then he told us that was a clue. The light dawned in our small brains! Something was happening along the sequence of operations that meant that no matter what numbers we started with, we all – including my father – ended with the same number. One of the operations along the line wiped out the difference in our mental number sequences. Knowing that, it didn't take us long to figure out how it was done.

My father's maths game is a loose analogy for the way 'inflation' theory helps solve some of the mysteries of the fine-tuning of the universe. Looking at the universe in the way my brother and sister and I looked at the maths game before we knew the secret, it seems the only way we could have a universe emerge as it is today would have been by means of fine-tuning to an incredible degree of exactitude when the age of the universe was an unimaginably small fraction of a second. With inflation theory we don't need such a specific beginning to produce the specific result we see today. In the maths game, the beginning could in fact have been *any* number my brother and sister and I could have come up with, and my father could have made the game end with any number at all. We would be over-optimistic to think that inflation theory allows us that much freedom when it comes to describing the initial state of the universe, but it does allow us some. What is the trick in the universe game that wipes out the differences among the many ways the universe might have started out, and allows it to arrive at what we find today?

Though there are currently several versions of inflation theory, Alan Guth of MIT gets credit for originating the idea in the late 1970s. Guth discovered a process which, at a time less than 10^{-30} seconds after the Big Bang, could have caused gravity to become a huge repulsive force. Gravity, of course, usually pulls matter back and slows the expansion of the universe. Instead, during a period lasting only an unimaginably small fraction of a second, it would have accelerated the expansion, causing violent, runaway inflation in the dimensions of the universe from a size smaller than a proton to about the size of a golf-ball. Inflation theorists think any imbalance between the expansive energy and the force of gravity would have been wiped out by this period of runaway inflation. The theory also helps with other fine-tuning.

To visualize the version of inflation theory which has the most to offer in this regard, we must first imagine the universe expanding a bit before the period of inflation begins. To make it easier for us, we cut down on the number of dimensions in this picture by returning once again to the balloon analogy. Inflate the balloon a little – to represent the expansion of the universe before the runaway inflationary period – then draw a tiny red dot on the surface. Attach the balloon to one of those machines which inflate balloons and turn the machine on at maximum force. The balloon (we must imagine it's strong enough not to explode) inflates to something incredibly large. The tiny red dot itself becomes an enormous area. One interpretation of inflation theory asks us to think of that red area, rather than the whole balloon, as the entire observable universe that we know today. In other words, what we are able to observe is only a very small fraction of all there is.

Let us instead draw dots all over the balloon, perhaps an infinite number of dots. After we inflate the balloon each 'dot' will represent something equal in size to our present-day observable universe. Or perhaps not – and here we receive some of the help we need with the 'universe-as-a-put-up-job' problem. Perhaps the universe before the period of inflation was in a chaotic condition. We can imagine this as something like the surface of the ocean. It would be ridiculous to talk about *the* initial state. We can find all sorts of initial states on that surface, depending upon which speck of it we examine (see Figure 5.1). Because local conditions for each of the dots on the balloon surface would be different, each dot would respond differently to the gravitational repulsive force when it came. Some would not have the right properties to respond at all. But the theory tells us that, when the inflationary era ended, we would find that in any dot that had inflated, the force of gravity (now working in a way more familiar to us) and the repulsive force resulting from the original Big Bang explosion would be balanced in the way we now observe in our universe. In only a few dots would the other constants be so miraculously adjusted for life to evolve. Perhaps in only one of them. If so, then our visible universe is that one dot.

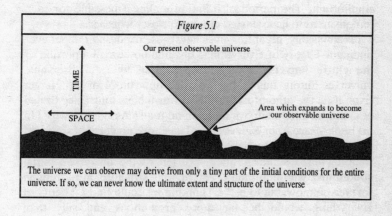

Figure 5.1

Our present observable universe

TIME

SPACE

Area which expands to become our observable universe

The universe we can observe may derive from only a tiny part of the initial conditions for the entire universe. If so, we can never know the ultimate extent and structure of the universe

Russian physicist Andrei Linde has suggested an extension of this theory. In Linde's picture each microscopic region (each 'dot') that inflates is in turn made up of microscopic sub-regions, which inflate and are in turn made up of microscopic sub-regions – and so forth ad infinitum – an eternal inflationary universe scheme.

Inflation theory helps a little to dispel the gloom expressed by Hawking when he said that the anthropic principle was 'a negation of all our hopes of understanding the underlying order of the universe'. At least we can invoke the 'weak' anthropic principle rather than have to call upon the 'strong' version. (The 'weak' anthropic principle, you'll remember, presumes that somewhere in the universe at some time there is a place for us.) It isn't just blind 'luck' that we have this universe. It seems almost inevitable (inflation theorists cannot yet tell us how nearly inevitable) that some tiny portion of the pre-inflation map of the universe would have had precisely the fine-tuning that would lead to conditions right for us.

Inflation theory also suggests a solution to what is known as the 'horizon' problem in Big Bang theory. This is the problem of why the universe is so uniform on the largest levels, in areas so remote from each other that it seems radiation could not have passed from one to the other even at the earliest moments. Yet the intensity of radiation is so close to the same in those remote areas that it seems they must have exchanged energy and come to equilibrium. The period of inflation makes it possible for such remote areas to have started out much closer together.

Paradoxically, all of this gives new cause for despair about ever finding a Theory of Everything. Imagine how small a portion of the entire universe may be represented by our observable universe. Surely this is the point-of-view problem writ large. Even if all the other dots besides our own have long since fizzled out, can we ever claim we understand it all? As Barrow wrote in his book *Theories of Everything*, 'The entire universe of stars and galaxies on view to us, on this hypothesis, is but the reflection of a minute, perhaps infinitesimal, portion of the universe's initial conditions, whose ultimate extent and structure must remain forever unknowable to us.'[18] Thinking we could find a Theory of Everything would be far more pretentious and silly than expecting to extrapolate the entire geography of the earth from an area the size of a pin-point three miles east of Colchester. No matter how clever our theories, we are likely to be prisoners of an extremely limited point of view – flat-earthers of the worst sort.

Inflation theory assumes a situation in which the laws of physics cause inflation to occur, but it is not an explanation of how the laws came to be what they are. We are left, again, groping for a more fundamental explanation.

BABY UNIVERSES TO THE RESCUE!

In Chapter 4 we learned that wormhole and baby-universe theory suggests a way to cut the Gordian knot of singularities. Now we're going to find that it also takes a whack at this second Gordian knot, the anthropic principle.

The 'cosmological constant' is one of the values that seem to require fine-tuning at the beginning of the universe. You may recall from Chapter 4 that Einstein theorized about something called the 'cosmological constant' which would offset the action of gravity in his theory, allowing the universe to remain static. Physicists now use the term to refer to the energy density of the vacuum. Common sense says there shouldn't be any energy in a vacuum at all, but as we saw in Chapter 4, the uncertainty principle doesn't allow empty space to be empty.

Just as the uncertainty principle rules out the possibility of measuring simultaneously the precise momentum and the precise position of a particle, it also rules out the possibility of measuring simultaneously the value of a field and the rate at which that field is changing over time. The more precisely we try to measure one, the fuzzier the other measurement becomes. Zero is a very precise measurement, and measurement of two zeros simultaneously is therefore out of the question.

Instead of empty space, there is a continuous fluctuation in the value of all fields, a wobbling a bit toward the positive and negative sides of zero so as not to *be* zero. The upshot is that empty space instead of being empty must teem with energy. The energy density of the vacuum – the cosmological constant – ought to be enormous.

Meanwhile, the Theory of Relativity tells us that the presence of matter or energy causes spacetime to curve, or warp. This proposition has been tested and we have observed this phenomenon. The paths light travels from distant stars are indeed warped as they pass near our sun, just as the theory predicts. The photons (messenger particles of the electromagnetic force) are pulled slightly off track as they pass a massive body – in the same sense in which you would be 'pulled off track' by a warp in a board-walk if you were skateboarding and made no adjustment to compensate.

Putting two and two together: We know the presence of matter/energy causes spacetime to curve; the more matter/energy, the greater the curving. We know the vacuum seethes with energy. The curving effect of that energy (if the cosmological constant is negative) is sufficient to curl the universe up into the size of a small ball, or (if the cosmological constant is positive) to have driven the expansion of the universe in such a way that galaxies couldn't have formed. Clearly neither has happened. Both quantum mechanics and relativity have served us magnificently in many theoretical and practical ways. We have overwhelming reason to believe that both are superbly reliable theories, and yet, allowed to work together, they serve up this nonsense. This is not one of the beauty spots of physics.

Fortunately for us, the value of the cosmological constant is observed to be near zero. How can that possibly be so, when such reliable theory tells us it should be enormous? Do all the positives and negatives there in the vacuum really cancel one another out that effectively? This is highly unlikely. Physicist Sidney Coleman of Harvard has this to say: 'Zero is a suspicious number. Imagine that over a ten-year period you spend millions of dollars without looking at your salary, and when you finally compare what you spent and what you earned, they balance out to the penny.'[19] For the cosmological constant to balance out to as near zero as it does is even less likely. That value would have had to be 'set' in the very early universe with a precision that defies understanding.

How can wormholes help solve this puzzle? Coleman explains it like this. Imagine the birth of a universe, as we did in Chapter 4 – a 'baby' branching off from an existing universe. Remember that according to the theory there may be plenty of universes around – some more enormous than ours is today, others unimaginably smaller than an atom, and all sizes in between. The new-born universe must copy its cosmological-constant value from one of these other universes through a wormhole attachment – 'inherit' it, you might say. It is of little importance to a human infant whether it inherits a talent for mathematics; it becomes important only when the infant grows larger. It isn't

172

important to a baby universe whether it 'inherits' a cosmological constant value near zero or one which would curl it up into a small ball. Its cosmological constant value won't even be measurable until it's quite a bit more grown up. However, with all those myriad assorted sizes of universes around, the infant has a far better statistical chance of inheriting its cosmological-constant value through wormhole attachments with large, cooler universes of the sort possible only when all those positives and negatives in the vacuum do cancel out to near zero. Coleman calculated the probability of a universe (in wormhole theory) being a universe where the cosmological constant is near zero: our sort of universe. He found that any *other* sort of universe would be highly unlikely.

Wormholes and baby universes are not able to work such magic with other constants, such as particle masses. Coleman and Hawking tell us that if we knew the map for the entire labyrinth of wormholes and universes, these values too might be explained, but of course we have no way of studying that map. All we can conclude is that, if these theorists are right, the reason why these constants of nature are knowable only by observation is that they arise from a situation in which some randomness plays a part – leaving us at best calculating probabilities, not with exact predictions.

NOT THE ETHER AGAIN!

For most of the twentieth century, physics students have been taught that the 'ether' – an invisible medium which was once thought to pervade all space – does not exist. The presence of such a medium seemed necessary in the mid-nineteenth century for explaining how light waves could be carried through space, and in Newton's time (though Newton himself rejected the idea) for explaining how bodies separated by space can exert a gravitational pull on one another. Einstein, our physics textbooks tell us, demolished the ether for ever. There is a genuine void out there – a vacuum state.

You will of course have noticed that this is not the picture we get these days from physics theory. Empty space very likely seethes with energy. Currently, theorists are gaining a deeper understanding of one kind of energy which may pervade all of space. It is the Higgs field.

The Higgs field is thought to be a sea of energy in which we and all other things in the universe are swimming. However, the only direct effect the Higgs field has is the existence of that attribute we call 'mass'. Savas Dimopoulos of Stanford University and CERN, Lawrence Hall of the University of California Berkeley, and Stuart Raby of Ohio State University hope that by understanding how the Higgs field interacts with certain particles they may be able to derive the values of some particle masses which up to now have been part of the unexplained fine-tuning of nature.

Dimopoulos tells us that we can imagine the Higgs field as a viscous medium – honey perhaps – filling all of space. A frisbee moving through the honey picks up sticky stuff and becomes heavier. A coin moving through it picks up less and gains only a little weight; a pinhead, even less. In the very early universe, particles moved through the Higgs field, this sea of energy, in a way analogous to the way frisbees, coins, and pinheads would move through the honey. Those particles we have come to know as the third family of leptons and quarks (see Figure 5.2), the heaviest family, were like the frisbee. They were particles which gained a lot of weight. The second family were like the coin. They didn't gain as much weight as the third family. The first family (which we know now as the electron and the two quarks that make up protons and neutrons) were like the pinhead. They gained even less weight. There is also a sense in which each of the lighter families absconded with weight from the next heavier – a slight case of cannibalism.

All of this weight-gaining resulted in six kinds of quarks and six kinds of leptons, grouped in the three families (Figure 5.2). The third and lightest family, including the electron and the 'up' and 'down' quark (which make up protons and neutrons), are still familiar inhabitants of the universe on the particle level. The other two families, including four leptons and the 'charm', 'strange', 'top', and 'bottom' quarks, not so. They existed in the first split second after the Big Bang, but we have managed to produce them only in particle accelerators.

Studying all of this weight adjustment, and the mathematical

Figure 5.2

THREE FAMILIES OF PARTICLES THAT SEEM TO MAKE UP ALL MATTER.
EACH FAMILY CONSISTS OF TWO LEPTONS AND TWO QUARKS

	LEPTONS	QUARKS
LIGHTEST FAMILY Particles found in ordinary matter:	electron neutrino	up
	electron	down
HEAVIER (More massive) particles:	muon neutrino	charm
	muon	strange
HEAVIEST FAMILY:		

symmetries the theory indicates should have resulted among the three families, Dimopoulos, Hall, and Raby have derived seven particle masses which were previously derivable only from observation. If their work continues to produce such results, and if it can be confirmed by experiment, the assault has begun in earnest on constants of nature which have been arbitrary elements in all previous theories.

There is hope that in the late 1990s and early 2000s the Large Hadron Collider (LHC) at CERN will have the capability of producing the Higgs particle. Experimenters will try to create a disturbance in the medium (the Higgs energy field), like a wave in the ocean: 'Kick it a little,' as Dimopoulos puts it. That wave would be the Higgs particle. It will not show up as a track on a photographic plate, but, after an extremely short life, it will decay into other particles that will leave tracks. Experimenters hope to see a shower of debris showing evidence of such particles as quarks and maybe even W and Z particles – a shower they believe to be characteristic of the decay of a Higgs particle. That discovery would not only lend support to Dimopoulos, Hall, and Raby's theory, but also help us to understand the symmetry-breaking in the electroweak theory.

Of course, though the Higgs particle and the Higgs field are necessary to the theory proposed by Dimopoulos, Hall, and Raby, the discovery of the Higgs particle would not in itself serve as confirmation of their theory – which predicts how the Higgs particle interacts with quarks and leptons. Far more meaningful in that regard was the discovery of evidence for the top quark by an international team of scientists at Fermilab in Chicago, announced in April, 1994, indicating that the mass of this quark is close to what Dimopoulos, Hall, and Raby's theory predicts.

The work Dimopoulos, Hall, and Raby are doing may help us predict some of the constants of nature, and perhaps tell us how and why it is that matter has the mass it does, but will it solve the mystery of why some of these relationships should appear to be tuned to allow for our existence? According to Dimopoulos, the first task is the practical one of finding out which symmetry actually works, which is the right symmetry. Beyond that we might ask whether there are underlying reasons why this symmetry and not another should be the one to apply in our universe. But the answer to that question still might not tell us why those values so crucial to the existence of life as we know it – the masses of the electron and the lightest family of quarks – should happen to be the right values to allow this life to exist.

Let's pause to see where this discussion about purpose and design has taken us. We have been considering a belief in a God who not only created the universe but had a purpose in creating it. Part of this purpose was the creation of beings capable of searching for God and responding to God. We asked whether that belief would conflict with modern science.

First, we saw that by the process of evolution the world we observe today, which seems so richly and miraculously designed, could have evolved without requiring a designer. The 'argument from design' is not a compelling argument for the existence of a designer God. However, we also learned that evolution does not rule out belief in a designer God. If God created the process of evolution and the ground rules by which it works, he set up a system on this earth which most evolutionists believe would inevitably at some future time produce creatures with sufficient intelligence and self-awareness to respond to God. God would not have had to interfere with the normal working out of the process in order to end up with such creatures. Thus one is free to believe that God invented the process of evolution in order to

176

create, and that he intended creation to include such creatures as ourselves.

However, evolutionists tell us that evolution, DNA, and the way it operates also do not need a creator or designer. The odds of those arising spontaneously and evolving to their present forms are also good enough to explain them. Once again, the argument backfires against those who would use it to banish God from the picture. The very same arguments which allow us to conclude that the odds are good enough also show that a designer, again without interfering or fiddling the numbers, could have allowed things to take their course and ended up with what he had intended from the beginning – creatures who would respond to him. We can come to no other conclusion based on the statistical situation presented by Dawkins, although Dawkins would scoff at the notion of attributing any role in all this to God. It is, in fact, hard to imagine a cleverer way of producing these creatures while at the same time allowing the freedom and contingency that seem such an important ingredient in this universe. According to the strongest version of modern Darwinism, all God really needed to do was to invent the laws of statistics. On the other hand, the laws of statistics may be dictated by mathematical consistency. And mathematical consistency, not God, may have been the First Cause – what just IS. We are back once again to our stand-off.

So far we find nothing profoundly inconsistent between belief in the God with a purpose and what we learn from science. Different interpretations and speculations are possible, but we must admit, whichever interpretation is ours, that we cannot prove that the other interpretation is wrong. Dawkins does not claim to have proved there is no God. He only claims to have shown that we can't use the argument from design to indicate that there *is*. By the same token, the arguments which show that evolution would have been a clever and almost fool-proof way for God to set things up do not prove that there is a God who did this. They only show that we can't use the process of evolution to prove that there is no such God.

The argument from design receives some support from the fact that only by incredible fine-tuning at the moment of creation could we have ended up with a universe in which living beings could have arisen. The anthropic principle is an alternative explanation, but that principle is unsatisfactory to many scientists as well as to those who would like to see a role for God in this fine-tuning. Physics has proposed some theories that offer other suggestions for how the universe could appear to be a 'put-up job' without really being one, and the effort to understand the constants of nature continues. No theory to date is able to explain *all* the fine-tuning, and none rules out the possibility of a creator God as the First Cause – setting up the context necessary for the theories to work. As we've seen in previous chapters, much hangs on the question of whether mathematical and logical consistency confines us to one theory, and whether mathematical and logical consistency requires an inventor – whether it could be different from what it is.

THE LONGING OF JOHANNES KEPLER

A God who invented mathematical consistency (which is also of course necessary for the theories of inflation, wormholes and baby universes, and the Higgs field), and then sits back expectantly and waits for the proper sort of universe to show up and life to appear and sentient beings to evolve who will respond to him, is a safe God to believe in if you don't want to come to blows with science. I have stated that glibly, in a way which doesn't reflect the depth of thought and reverence with which some have arrived at that belief, nor the wealth of personal variations. There are many possible implications of the words 'responding to God', and many shades of belief about how God responds to us.

Perhaps God knows each one of us as an individual. Perhaps it matters to God what happens to us. Perhaps God offers hope of life after death (maybe with reward or punishment). It's possible God might do all this and yet maintain a strict policy of hands-off in the universe.

When my daughter Caitlin was five years old, my husband and I took her to London, where we watched the pomp and circumstance as the Queen rode in her carriage to the opening of Parliament. As the Queen's carriage passed us, she waved and nodded directly at Caitlin. Later, when asked whether she had seen the Queen, Caitlin responded, 'The Queen saw *me*.'

Similarly, how far more significant than 'I know God' is 'God knows me.'

It seems so unlikely – much less likely than the Queen of England taking a personal interest in Caitlin. Hawking has said, 'We are such insignificant creatures on a minor planet of a very average star in the outer suburbs of one of a hundred thousand million galaxies. So it is difficult to believe in a God that would care about us or even notice our existence.'[20] Psalm 8 is the same reaction from a believer's point of view: 'When I consider your heavens, the work of your fingers, the moon and the stars, which you have set in place, what is man that you are mindful of him and the son of man that you care for him?'[21] What indeed!

The question isn't whether we are being too modest when we take this attitude, but whether this isn't a very limited, human idea of God. Would any God worth his salt be subject to size or distance constraints? We can't accept Hawking's or the psalmist's quibbles on these grounds as scientific evidence for the lack of a God who is aware of us.

This concept of God – loving us but maintaining a hands-off policy – has some explanatory power when it comes to the mind-shattering cruelty, blatant unfairness, and sheer absurdity in this universe. This is a God who refuses on principle to meddle in his creation once it's set up on the most basic level – who allows the normal process of health and disease and happiness and disaster to take its course. Some believe that 'hands-off' does not mean detachment, and that God's grief is grief beyond human capacity, for an infant born deformed or with severe mental retardation, for a child battered or starved to death, for a young man killed by accident by a shell from his own comrades in a war, for a genius trapped in a paralysed body – all situations which seem unconscionable if God *does* intervene in the universe. The picture accords very well with life on this earth as we normally experience it. Also, there's nothing about this belief that conflicts with science, though at the same time the question begins to loom large: 'Can you give me a good reason why I *should* believe such a God exists?' From the point of view of science, the greatest weakness of this belief lies in its unfalsifiability. What direct evidence could we possibly look for in this world to show whether or not it is correct?

What if we turn a few pages in the Bible and quote Psalm 46:

'God is my refuge and strength, an ever-present help in trouble. Therefore I will not fear, though the earth give way and the mountains fall into the heart of the sea, though its waters roar and foam and the mountains quake with their surging.'[22] 'Ever-present help' implies something more than the (perhaps) comforting knowledge that God is sympathizing from a distance and may compensate us in a later existence. If God is in any active way a refuge and strength, we no longer have a God with a strictly hands-off policy.

Johannes Kepler was a scientist who believed in God. One of the most significant contributions anyone has made to our understanding of the universe was his suggestion, which followed soon after Copernicus' sun-centred astronomy, that the planets move in ellipses rather than in circles. It was a radical idea, partly because it had religious and philosophical implications. In Kepler's own way of expressing it he had, by suggesting these elliptical orbits, 'laid a monstrous egg'.[23] For centuries philosophers, scientists, and theologians had considered the circle as a manifestation of perfection. Ellipses certainly looked like a move in the direction away from perfection and beauty. But Kepler showed that ellipses explain the movement of the heavens in a much simpler, more elegant way than circles do.

Kepler was a devout Christian who wrote part of the Protestant Prayer Book, but he had no qualms about relinquishing the belief that God, being perfect, prefers circles. It was Kepler, writing about his discovery of elliptical orbits, who exclaimed 'O God, I am thinking thy thoughts after Thee.'[24] It was also Kepler who wrote 'There is nothing I want to find out and long to know with greater urgency than this. Can I find God, whom I can almost grasp with my own hands in looking at the universe, also in myself?'[25]

It is undeniably a tremendous leap from belief in the hands-off God to belief in a God who makes himself an active part of our universe – not just a part of our universe in a general way through the implacable and eternal laws of physics, but 'a part of myself'. The message of the Old and New Testaments is that the Spirit of God communicates with people and even lives within them. Through those who allow this to occur, God significantly affects the course of events. The other side of the coin is that when we harm or neglect people we may be harming or neglecting God.

180

For those who believe that God is beyond us and beyond the universe but also near us and in communication, it becomes not unthinkable for prayers to take such forms as 'Give me strength to endure this ordeal', 'Guide my hands as a surgeon', 'Show me what you want me to do', 'Help me overcome this addiction', 'Give me the ability to love this unlovable person'. Nor do such prayers have to be only for one's self. One might reasonably expect the response to be anything from a vague influence to a distinct inner voice or an almost irresistible directing force – perhaps sometimes not a response to one's liking, but a definite response nonetheless. Those who believe in this God believe they can wrestle in their minds with him, expecting real input and opposition. If human beings are not just beings who respond to God but channels for God's power and influence, that gives God a potentially significant handle in this world. Anything is possible that can be accomplished through consenting human agents. Belief in such a God is consistent with the assumption of free will, if God intervenes only upon invitation and one remains at liberty to disobey God's instructions. Depending upon how much our minds influence our health and physical bodies (a still hazy area of science), it might be possible to extend the argument to allow for physical healing in answer to prayer without coming into conflict with any of the physical or biological laws that we know about.

There is a widespread assumption that belief in a God who intervenes in the universe by using human beings as his intermediaries would be difficult, even impossible, to save if science were able to explain consciousness, self-awareness, intellect, personality, emotion, intuition, aesthetics, inspiration, and belief, in terms of physical and biological processes, perhaps by demonstrating that our brains and nervous systems are supercomplex computers, their hardware and programming the product of evolution. Science to date has, of course, not managed to arrive at any such explanation for our minds, and there remain great uncertainty and disagreement about whether we can ever expect to have one.

Those who believe science will succeed in explaining the mind in physical terms point to promising advances in psychology, neurological science, theories about the way the mind has evolved, as well as in the rapidly developing fields of artificial intelligence and artificial life. In this book we have seen numerous evolutionary explanations for what is mysterious about our minds: the way our mathematics parallels nature, our ability and proclivity to philosophize and to ask questions about God, our notions of 'good' and 'beauty', the fact that we find the universe 'rational', even our definition of rationality. We understand that our emotions, our tastes and preferences, our reactions to one another and to the circumstances of our lives, are at least in part explained by genetics and chemistry. Artificial-intelligence and artificial-life experts hope consciousness may eventually be simulated to the extent that the simulation really *is* consciousness, indistinguishable from our own.

Those who think instead that science will *not* succeed in a complete explanation for our minds point to other evidence currently emerging which indicates that the mind will never be entirely explained as a super-complex computer. Even our mathematics goes beyond computability. Penrose wrote in the conclusion to *The Emperor's New Mind:*

> I have presented many arguments intending to show the untenability of the viewpoint – apparently rather prevalent in current philosophizing – that our thinking is basically the same as the action of some very complicated computer . . . there must indeed be something essential that is missing from any purely computational picture. Yet I hold also to the hope that it is through science and mathematics that some profound advances in the understanding of mind must eventually come to light.[26]

Pippard writes:

> Too many physicists (and others) take for granted that in due course an explanation will be found of conscious mind in terms of the material operations of the brain. This is to put the cart before the horse – it is through our minds that we know of the brain, and we are more likely to find how they are related by concentrating on the fundamental thing (conscious knowledge) rather than its derivative (material brain).[27]

Those who remain optimistic about finding a complete scientific explanation do so not because we are near to finding such an explanation, but on grounds of personal preference and the assumption that science and mathematics are irresistible forces to which all barriers must eventually fall. The sceptics are agnostics (such as Penrose and Pippard) and atheists as well as some who believe in God, so we cannot assume that such scepticism necessarily reflects religious bias.

Clearly the verdict is not yet in. However, it is a notorious pitfall to save religious belief by talking about what science hasn't been able to answer, or looks unlikely to explain. Need we do that in order to save belief in a God who influences the world through human agents? Arguably not. If our minds are explained as super-computers, what relevance does that have for whether one computer communicates with other computers, whether one human communicates with another human, or whether a human communicates with God? Furthermore, can any computer ever be certain it knows the source of all its hardware, programming, and input? Can any computer ever be certain it knows the source of all its fellow computers' input? The 'profound advances in the understanding of mind' Penrose hopes we will find through science and mathematics do not promise to be a complete explanation, and we are far from knowing even what sort of advances we might be looking for.

Saving belief in God by talking about what science hasn't been able to explain, or looks unlikely to explain, *is* skating on thin ice. It is disparagingly known, among modern theologians, as God-of-the-Gaps theology. On the other hand it is no more intellectually viable to save *un*belief solely on the assumption and hope that science will inevitably be able eventually to explain everything. We've allowed ourselves a stand-off on the grounds of 'It remains a mystery' in the First Cause contest between God, mathematical and logical consistency, and the universe, because *no-one* can show that science as we know it will ever objectively resolve that chicken-and-egg puzzle in a definitive way. At present we also have no other choice but to allow a stand-off regarding an explanation for the human mind.

Suppose God has other means besides human minds and human agents through which to act in this universe. Suppose God suspends the normal chain of cause and effect in answer to prayers or for reasons entirely his own. Suppose God occasionally or even often sets aside the physical and biological laws of the universe as we understand them. There are shades and degrees of belief among religious persons concerning how much God does interfere and in what ways. Suppose we even go so far as to believe the Old Testament story about God stopping the sun so that Joshua and his army would have sufficient daylight to win a battle?

The leap to belief in an active, intervening God is a hazardous one, because it makes specific predictions about events we ought to be observing if they happen. It is, of all forms of belief, the most vulnerable and potentially falsifiable.

This is not a book designed to pit science against religion in the conventional way and come up with a winner. But as this chapter has progressed, we have pursued the questions with which we began in ever more challenging detail. Is it possible to believe wholeheartedly, without compromise, in both orthodox late twentieth-century science and God – or would such simultaneous belief be double-think or other intellectual dishonesty? If it is impossible, what belief about God or what scientific knowledge makes it impossible?

We've insisted on taking science and religion at face value, not looking for a reconciliation or ways one approach must be torn down or compromised in order to allow for the other. We've invited science and religion to come out on this quest in full armour and regalia, all banners flying – and for the occasion we've suspended any suspicion that either or both are riding imaginary horses. Yet the tournament hasn't actually produced a serious contest. The participants have ridden past each other, their lances landing glancing blows at best. Haven't we already gone far enough by positing a God who might actively intervene in human minds? Must we now risk a bloody battle by considering a God who intervenes actively in the universe?

6

THE GOD OF ABRAHAM AND JESUS

Why does the universe bother to exist? One can of course define God as
the answer to the question, but that does not advance one much unless
one accepts the other connotations that are usually attached to the word
'God'.

STEPHEN HAWKING[1]

'THERE IS NOTHING NEW UNDER THE SUN,'[2] SAYS THE BOOK OF
Ecclesiastes.

Several years ago I overheard a conversation between my
brother, who is an agnostic mathematical physicist, and a friend
of ours, a well-educated musician who believes in God. She
described to him an experience which she couldn't think how to
explain except as a miracle. My brother laughed. This response
seemed uncharacteristically rude. It would have been more his
way to mutter something non-committal or apologize politely for
having to disagree with her. But he insisted it was the appropriate
response, 'for two reasons, actually. One – it sounds like total
foolishness to me. Two – if it isn't total foolishness, it's a first-rate
reason to laugh.'

Steven Spielberg's science-fiction film *Close Encounters of the
Third Kind*, about the arrival of aliens on earth, expressed in a
whimsical manner the mixed feelings of fear, scepticism, awe,
hope, freedom, and incipient laughter at ourselves and our

mundane expectations which human beings experience when they allow themselves to believe the words 'And now for something completely different!'[3] The promise is seldom, if ever, kept. 'There is nothing new under the sun' is more our common experience. Part of the appeal of the Christmas story is its message – so different from Ecclesiastes – regarding the establishment of a new relationship between God and humanity, a new beginning with potential echoes in the lives of individuals. How many persons long for a fresh start, a sense that 'anything is possible'!

Surely this longing is often partly a nostalgia for a time in childhood when all possibilities did seem to be open. Could it be that we, even as jaded adults, might still find the hidden gate into the Secret Garden or the unlikely passage through the wardrobe into Narnia? Is there any escape route from the ordinary and the tediously predictable?

Escape to what? *Would* it be wonderful? Those who venture beyond the rational and the predictable don't always return enamoured of what they find. It isn't always attractive, nor is it Plato's world of forms. It is *The Rhyme of the Ancient Mariner*, *Death in Venice*, *Heart of Darkness*, *The Exorcist*, *Fanny and Alexander*, *Equus*, the Book of Job, or the old, uncensored versions of the fairy tales that some don't read to their children nowadays.

Science assures us it is all explainable. If there *is* country beyond the gate or the wardrobe, we need only venture forth and build the wall of our understanding further out – which we can do, given time and ingenuity. Art, literature, music, and, some would add, experience tempt us to suspect we can never extend the wall far enough, that there is reality which we can never hope to explain and understand by means of human reason.

Belief in the God of the Bible is much more radical than this belief that something lies outside the boundaries of our presnt rationality, or the belief that this something occasionally crosses the boundaries and invades our everyday territory, or that a few of us – saints? madmen? – occasionally venture or blunder out beyond what we normally think of as normal and rational. It's a belief that God is the God of all of it, that both sides of the boundary are real and rational and that they interact continually. If anything is imaginary, it's the boundary.

How does *that* wash with our science?

Pose the question 'Can I believe strongly in the scientific view of the universe and at the same time believe in a God who is involved continuously in events in the universe?' and you will get a variety of answers:

1. 'Best to ignore all that religious hocus-pocus. There's no scientific evidence for any of it. It's all superstition. Stick to what you can learn from science.'

When we offer that advice, we imply that we have asked and received an answer to the question we imagined our hypothetical alien asking in Chapter 5. Despairing of any evidence to disprove the existence of God, the alien asked: 'Can you show me evidence that there *is* a God? Something other than the fact that nobody can prove there *isn't*?' That question of evidence looms very large when we talk about a God who not only created the universe but is actively involved in it. If this God exists, we certainly ought to see some clear signs of his activity.

2. 'You can believe in God, but believing that God is involved in the universe in the way the Bible claims he is – that's unsatisfactory to the scientist or anyone else who understands and appreciates the fundamental rationality of the universe. If there is a God, God is rational and faithful. How could such a God regularly break his own laws? After all, for those who do believe in God, the rationality of the universe is one of the best indications of the nature of God's mind. How could God be false to his own nature?'

That is an argument we are as likely to hear from believers as from atheists and agnostics – any with a strong commitment to a rational universe which operates by laws whose beauty and elegance are cause for celebration. There is even a Christian hymn which goes: 'Laws which never shall be broken for their guidance He hath made.' Having read thus far in this book you ought to have some sympathy and understanding for that point of view, particularly when you hear it from a scientist. We can hardly do science without a belief that there is dependable pattern and order to the universe. The way what is confusing or unknown eventually becomes clear to us through exercise of our intellects, the way our thinking can predict the reality of the universe – mathematics matches nature, theories are confirmed by observation – all of this argues persuasively that the universe is rational and that this rationality somehow parallels our own. The notion that a superior might barge in and mess with the laws we have discovered challenges our deepest faith in science, calls into question the principles on which that faith is based, and makes a mockery of the truths science has uncovered.

3. 'Why should it be any more "breaking laws", if God intervenes in the universe in ways that seem to undermine its predictability than it is when traffic regulations are set aside to allow passage for emergency vehicles or to permit a parade to take place? That's not breaking laws; the basic laws of the country are not being violated. Let's replace this excessive legalism with an attitude that allows for common-sense flexibility.' This argument insists that 'rational' doesn't have to mean 'legalistic' or 'deterministic' and we oughtn't to confuse the terms.

4. 'Laws of science? Which laws of science? The ones I learned at school in the sixties? The laws of science as of last October? The laws of science in the year 3000? Science is a shifting body of knowledge. The monolithic, implacable 'laws of science' which God is supposed either to break or not to break – which ones are they? Where are we going to take our slice and say "That's it"?' Any exercise in which we set religious belief alongside scientific knowledge may flounder on a serious point-of-view problem – point of view prescribed by what high-profile scientists have concluded as of today or, more likely, as of the most recent time you or I read a science magazine.

5. 'If you want to believe in a God who is involved in the universe, you're going to run into a much more profound problem than a few broken laws of physics. Nothing is more difficult to accept than the apparent irrationality and arbitrariness of a God who dabbles in the universe, but who does not do so more often, and who seems to play favourites. This is a God who helps a matron in Kensington find her lost poodle but allows thousands of children to starve in Somalia. Any decent human could do better on an off day.'

Here, certainly, is the most devastating argument against belief in a God who is involved in the world. Although this is not strictly an area of contention between science and belief, we can't talk about a rational universe and a rational God and fail to deal with it.

6. 'God isn't breaking any laws when he intervenes, nor is

God's activity in this world senseless and inconsistent. Science may be the jewel in the crown of human intellectual endeavour, and our concepts of goodness and fairness may largely come from standards set by God, but let's not fool ourselves into thinking that what we call a scientific point of view is THE last word in rational, objective points of view.' This reply suggests we return to the metaphor we used in Chapter 3 and think of science as capable of drawing a picture of the room that allows us to find our way around adequately, sometimes even brilliantly, while never being able to show us all there is in the room or help us find our way in all circumstances. We said before that science must choose a few out of an infinite number of ways of ordering and explaining reality, but we also implied that science could potentially have the entire range of choices at its disposal. Now we are suggesting that there might be some views of the room which science could never discover – and they might be by far and away the most significant ways of looking at the room.

7. 'We have the universe we have; and we have the God we have. Let's quit arguing about what *ought* to be possible, what *ought* to be allowed, what contradicts what, and focus instead on what *is* happening – and take it from there. That's a much more scientific approach.' With that argument we return to the question of evidence. Is there hard evidence for the existence and the active involvement of God? Does such evidence, if any, conflict with scientific evidence? If so, then certainly arguments about what can be or ought to be should take second place to finding ways of resolving or living with *real* – not merely hypothetical – contradictions.

In this chapter and the next, we'll let these seven arguments speak to one another and see where they lead us.

THE LAW-BREAKER

If our contemporary knowledge of science can be termed 'laws', then judged in the light of those laws the God of the Bible is an outlaw. If God made the laws, that may give him all the more right to break them and be 'a law unto himself', but it certainly prompts some questions. Why didn't an all-knowing and all-powerful God make the laws perfect in the first place so that he wouldn't have to fiddle around with them later? Was it so that we humans could have free will and God could still manage to save us from the worst possible consequences of our choices? Was it so that the natural universe could have contingency but, again, God could manage to save it from the worst possible consequences? Why should God, who presumably could have created a perfect universe, have chosen to create a universe in which those worst possible consequences were even possible? Does God change his mind? What sort of messy creation is this that requires this perpetual dabbling and tinkering?

There are, as we've said, believers in God who insist that any dabbling and tinkering and breaking of natural laws is at odds not only with scientific knowledge but with the concept of a rational God who deals faithfully and intelligently with his universe. Physics Nobel laureate Sir Nevill Mott writes:

> I believe that the laws of physics and chemistry are not broken; water is not turned into wine and a body is not removed miraculously from a tomb. I must believe, if God is omnipotent, that He could do these things if He wanted to, but I cannot worship or respect a God who would want to. Such a God is to me a tribal god, showing off to his followers that he has that kind of power . . . My assumption is that God relates to men and women who seek him, and that He works within natural law.[4]

On the other hand, among those people who are troubled with the notion of a God who breaks laws, there are some who nevertheless believe that God does indeed intervene in this universe. Some of these even claim they have had personal experience of this intervention. For them the problem isn't an abstract one, but one of resolving in their own minds what looks to them like a worrisome contradiction. Since it may not be immediately obvious exactly what manner of contradiction we are dealing with, some further explanation is in order.

Many, though not all, events which the Bible treats as divine intervention do not contradict science or break any scientific laws. In fact even the definition of a miraculous event as something which can't be explained by science, or which contradicts what we know from science, is not a good one. If you wish to test this, describe a miracle to a scientist who doesn't believe in miracles. The reaction may be laughter, as in the case of my brother, but you are more likely to hear an explanation that will show how this miracle could work within the scientific scheme of things. It's a hallucination, or a coincidence, or an instance in which something that's very improbable but nevertheless possible has happened, or an example of how amazingly our minds influence the physical processes of our bodies. If you want to go on believing that the event is a miracle, that's your choice, but you have to admit there's another possible explanation. The miraculous claim, then, is not that the event happened, but that it happened because God caused it to happen and the natural explanation isn't the complete explanation.

In such cases, in order to pinpoint the contradiction we're looking for, we have to pose a different question. Instead of asking whether the event in and of itself contradicts scientific knowledge or breaks the laws of nature, we have to ask whether the divine *explanation* contradicts scientific knowledge or breaks these laws. For example, rather than say the parting of the Red Sea didn't occur, because such an event contradicts the laws of nature or can't be explained by science, some have tried to account for it as a natural result of the same volcanic explosion which resulted in the island of Santorini. Whether that explanation has any validity is not at issue here. The point is that, even if we accepted it, the question would still remain whether it contradicts scientific knowledge to believe God was the instigator of *whatever* chain of events caused the Red Sea to part at the strategic moment for the Israelites to cross, and to close again when the Egyptians followed them. Finding a scientific explanation for a miracle does not mean we have found a reconciliation between science and belief.

Nevertheless there are those who find it acceptable to believe in a miracle that can be explained by science, but not in one that cannot. The argument is that God cannot or must not do things which could not happen purely as the result of natural processes. Whatever the value of that position, the line is not so easy to draw. For example, many who accept most miracles of healing baulk at believing that God stopped the sun for Joshua. Their reaction to that account is that it is simply unacceptable in the light of modern scientific knowledge. Is it?

For background, the story reads as follows in the Book of Joshua in the Bible:

> On the day the Lord gave the Amorites over to Israel, Joshua said to the Lord in the presence of Israel: 'O sun, stand still over Gibeon, O moon, over the Valley of Aijalon.' So the sun stood still, and the moon stopped, till the nation avenged itself on its enemies, as it is written in the Book of Jashar. The sun stopped in the middle of the sky and delayed going down about a full day. There has never been a day like it before or since, a day when the Lord listened to a man. Surely the Lord was fighting for Israel![5]

For the writer of that passage the most significant fact was not that the sun stopped, but that the Lord 'listened to a man'. We are far more sceptical about the sun stopping.

Even the most biblical-literalist among us is likely to accept the possibility that, in order to 'stop the sun', what God might actually have done was to stop the rotation of the earth. It's interesting to note that this is something like the approach Galileo took to the Joshua miracle, although in his explanation it was the *sun* which stopped rotating, not the earth. In Galileo's interpretation 'the sun stopped' was indeed the correct wording, but not in the sense we usually mean.

Galileo didn't claim that the event in the Book of Joshua wasn't a miracle. However, he sought to show that the biblical account described the experience and the appearance of what happened and didn't provide complete astronomical information about what caused this appearance. He used the story as an argument for his conclusion that the earth moved around the sun and not vice versa. His reasoning went something like this. If we look at this story from the point of view of having the earth orbiting the sun, then we see that what really happened was that the sun stopped all right, but not in the sense that it stopped moving across the sky. That was merely a secondary effect. The sun stopped rotating on its axis. Galileo had concluded that the sun's rotation drives the planets around. If the sun's rotation stopped, then the whole solar system would come to a halt until the sun started rotating again. This was not, of course, a purely natural explanation for the Joshua miracle. God would still have had to stop the rotation of the sun, and Galileo was willing to accept that. But his explanation seemed to him a far simpler means for God to achieve the desired results than all the tampering with the solar system that would have been necessary in order to achieve the same effect in the earth-centred Ptolemaic system.

Trying to find a way to get the effect described in Joshua from purely natural causes with no divine input whatsoever will pose a challenge for any scientist. We are almost inevitably forced to resort to the possibility that the model which makes sense of our observations of the way the solar system works fails to reflect a larger mathematical and physical description that we are not aware of at present. Perhaps in that description it would be easy to see why, in the course of many thousand years, the earth would once, for a few hours, apparently stop rotating (though rotating might not be the accurate term to use in this broader description); or why the sun and the moon would apparently stand still. Could the miracle be given a 'natural' explanation in that way? One would be hard put to say. In any case we haven't seen the end of our difficulties; they go beyond simply getting the miracle to happen.

Perhaps you're familiar with the Sherlock Holmes story which involves 'the curious incident of the dog in the night'.[6] The dog didn't bark. That was the curious incident. The wonder of the Joshua story is likewise not so much what happened as what didn't happen. Why were there no catastrophic climatic disturbances, no enormous destructive tides, no displacement of the earth's tectonic plates, no other gravitational consequences for the earth, the moon, or the rest of the solar system? If the planets stopped in their tracks (as Galileo hypothesized), they would plunge into the sun. We know today how intricately balanced and linked together all of this is – how sensitive even to small changes. Could any larger description possibly take care of all that too?

Perhaps a more promising possibility would be to abandon

attempts at a physical explanation and suggest instead that the psychological perception of time among those involved in the battle was temporarily altered. Possibly it was a mass hallucination.

If the sun had stood still during the modern Gulf War rather than in ancient Israel, scientists would certainly be scrambling to find the broader mathematical and theoretical description or to validate the psychological explanation. As it is, we have easier recourse to the claim that the Joshua event simply never happened at all.

That is not to say that we should scoff at the possibility of God performing this act for the sake of Israel. If one is going to believe in God at all, why should one not believe that God *could* do this with no more effort than a passing thought? As Mott said in the passage quoted above, the point is not that God couldn't have done it. The point is that, if it wasn't merely a psychological effect, this activity of God amounts to a massive intervention in the normal, law-abiding progression of events, at least in our corner of the universe. If we have strong scientific faith in a legalistic universe, or strong religious faith that a faithful God can never break his own laws, then we are going to find it difficult to believe this miracle really happened. If we witnessed it ourselves, we'd have a problem trusting our senses and our sanity. If many of us witnessed it, and we could show it wasn't merely a psychological effect, we would have to rethink so much of our science that we would hardly know where to begin. The simpler course might well be to concede that God did it, and let it go at that.

We encounter an even more significant clash between science and the Bible with the resurrection of Christ. Either Christ returned from the dead or he didn't. The New Testament insists that he did; and, if he did, we have an event which by the lights of present-day science breaks the laws.

What about the Genesis account of creation? Here the problem goes beyond dealing with what appears to be a miraculous event. What we have are two conflicting accounts of the creation of the universe and human beings – Genesis and science. It would certainly seem that both cannot be literally true.

As we proceed with this chapter we must be careful not to equate a belief that God interacts with and intervenes in the universe in ways similar to the events described in the Bible (a belief to be found across the full range of Judaism and Christianity, from conservative to liberal, and in other religions as well) with a belief that every event described in the Bible actually did occur precisely as described there. Such literalism is confined almost exclusively to the conservative end of the Christian spectrum; Jewish religious thought and tradition, for instance, have not accepted Genesis as a literal account. There are many who baulk at full biblical literalism who do not baulk at the notion of miracles and divine intervention in the universe. Most of this chapter does not juxtapose scientific knowledge and biblical literalism – though our discussion will often be relevant to that conflict, and we will not neglect it.

THE SOFT UNDERBELLY OF LEGALISM

One way those who believe both in divine intervention and strict scientific legalism attempt to deal with the problem of reconciling one with the other is to seek less obvious means, within the letter of the law, by which God could intervene. This is arguably an excessively legalistic approach. We become lawyers trying to show a jury that our client, God, may seem to have broken the law but, technically, hasn't. Nevertheless, before exploring some other approaches, let's dig in the law books of the universe to find out what we might come up with for a legal defence of God. That approach at least will take us through some very interesting science. We will confine ourselves to the mainstream, not resort to the more exotic fringes.

First, we must be clear about what we mean by the laws of nature, and why science can indeed be a continually shifting body of knowledge without our perception of those basic laws changing much more often than it does. When we speak of the laws of nature we do not mean laws that declare 'The sun may not stop', or 'The dead may not come back to life'. We mean laws that are much more fundamental than that, laws which underlie such statements. Specifically we mean laws which govern how things change and, by implication, how things are not allowed to change. Perhaps the greatest manifestation of symmetry and harmony in the universe is the fact that, although things do obviously change drastically over time, from place to place, and from situation to situation, the underlying laws, which govern how those changes occur, apparently do *not* change. Is this convincing support that our assumption of unity is correct, or is our assumption of unity leading us to a false impression that such symmetry exists? That is a question honesty forces us to ask. We can't answer it except to point to past experience in the search for these fundamental laws.

The search for a more fundamental law often begins with the discovery that something we have been regarding as fundamental and unchanging fails to hold under some circumstances. When this happens, our assumption of unity and symmetry kicks in and allows us to conclude that what we have been thinking of as a fundamental law is merely an approximation, and that we must now explore for a deeper underlying law which does not change. We haven't yet reached bed-rock.

We find many examples of this process of discovery in the history of science. For instance, the laws discovered by Newton hold except when movement approaches the speed of light or when gravity becomes enormously strong. Einstein's more fundamental description does not break down, as Newton's laws do, in these extreme situations, but Einstein's description in turn predicts singularities, and breaks down at a singularity. We assume that at absolute bed-rock there are laws which break down in no situations whatsoever. No-one pretends we've discovered that bed-rock in any area of science, but this doesn't mean that the laws we have discovered aren't extremely dependable in the situations we normally have to deal with.

The underlying unchanging laws, whatever they are, and the nearest approaches to them that we've found, do obviously allow a vast range of changes and events to occur – a vast range of behaviour and experience. As a loose analogy, we might compare them to the constitution of a country. A constitution certainly does not constrain most of one's behaviour and choices very severely or deterministically.

Getting back to our search for a legal way for God to intervene in the universe, a promising approach would be to look at the many instances in which the underlying laws of nature do not absolutely determine what happens, where they allow an element of chance, randomness, or choice. This approach isn't new. The notion that divine intervention might occur where there is an element of chance allowed people in the Bible to determine God's will by a throw of dice. The notion that it might *legally* occur thus, even if inflexible physical laws exist, has been around at least since the seventeenth century. We saw a contemporary example of this argument in our discussion of evolution. The underlying laws by which evolution occurs don't determine what mutations will appear. Presumably that would be a place where God could intervene without breaking any laws, no matter how much such intervention would offend anyone's sensibilities in other regards.

It isn't only God who may get an oar in at some of the interstices of chance and choice. We might say I 'meddle in the universe' if I decide to plant zinnias in my garden as opposed to begonias. No scientists tell me, when I show them the zinnia bed I planted myself, that they think it really must have happened through natural causes rather than my intervening in the universe (though they might say that, if they knew my usual lack of success as a gardener). However, a breeze could have blown my neighbour's zinnia seeds through the fence. Either explanation is plausible. But even if this planting is strictly my doing, no-one accuses me of having broken a law of nature. The laws of nature allowed me my choice.

I meddle in the universe if I am a botanist and I develop a new strain of zinnias. Nature also has ways of developing new strains of zinnias, but I perhaps find a way of doing it which nature never uses as far as we know. No laws have been broken.

However, someone would surely blow the whistle if I caused my zinnia bed to rise all by itself and float over the fence into my neighbour's garden. That's a different category of meddling in the universe. As far as we know, I couldn't do that without at least breaking the law of gravity – one of the laws which govern how things can change.

Short of performing such a wonder, it appears that within the confines of the laws of the universe there is immense leeway even for human beings such as myself to exercise the free will we assume we have. Though we don't understand how it is that the human mind is able to make such choices, we apparently do make them and exert quite a strong influence within our small part of the universe. The argument goes that God has left plenty of room

197

to manoeuvre within the laws he created, plenty of leeway not just for us but for himself, perhaps on a much grander scale and with infinitely greater understanding of the laws than that vouchsafed to you or me.

None of the following discussion of science which pinpoints areas where choice and chance occur, and asks whether they could be exploited by a being with infinite knowledge, should be taken to suggest that God is a physical agent like ourselves or an energy or force in a scientific sense. None of these arguments attempts to equate God with something (or some process) we have found in nature. None pretends to propose the means, physical or otherwise, by which God would intervene. They are only attempts to find out whether such intervention, should it occur, would be unfaithful to what some take to be God's own laws. If this degree of legalism bores you, the science I think will not.

We will be quick about the first suggestion, because it reiterates something we discussed in Chapter 5, that God might act in the universe by influencing the conscious and unconscious minds of human beings. If God works in that way, any choice I am allowed becomes by extension an opportunity for input from God. That wouldn't be breaking any laws of science that we know of, and it would allow God immense leeway. We suggest no contradiction to known scientific laws if we propose that God brought about many of the biblical miracles in this manner, those that didn't require the manipulation of the physical universe in ways independent of human intermediaries. This is not the same as saying that these miracles were hallucinations. The extent to which physical healing is connected with the mind is too poorly understood for us to rule out its taking place as a result of God's activity in our minds. Also, nothing in science indicates incontrovertibly that God would be tinkering in an illegal fashion with physical reality if God makes his presence and influence felt in visions and dreams, or if God communicates with us and answers prayers by offering guidance in the form of thoughts and inspiration.

However, in the Bible we are faced with a God who does also act in ways independent of human intermediaries. God causes floods and earthquakes and sends fire from heaven. God brings the dead back to life. This God's influence on the universe is not confined to his influence on and through human beings.

198

An obvious place to find a loophole of chance and choice might seem to be the quantum level of the universe. British neurophysiologist and Nobel laureate John Eccles thinks it is from the quantum level that the possibility arises for both human and divine choice. Cambridge physicist and theologian John Polkinghorne believes that God does indeed exercise choices in the universe. Polkinghorne doesn't discount miracles. However, he does not favour Eccles' interpretation which sees the quantum level as a loophole for God's activity. He points out that 'the aggregation of individually chance events at one level is liable to compose itself into a highly predictable pattern at a higher level . . . I am not saying that there are never circumstances in which quantum effects are amplified to have macroscopic consequences, only that they are unlikely by themselves to provide a sufficient basis for human or divine freedom.'[7] Polkinghorne also finds this type of activity a little too 'hole-and-corner', arbitrariness masked as faithfulness.

It is time for another look at the quantum level. In spite of the reputation the quantum level has for uncertainty and unpredictability, equations which govern that level are deterministic. That seems not possible, when they allow a particle to be in many places at once. Such results run counter to our normal experience. We have to stretch our imaginations to think of alternative positions coexisting in a way they cannot do in our everyday world, where a billiard ball or a planet is never in several places at once. The need for language to help us leads us to speak, for example, of electrons moving in a 'blur' around the nucleus rather than orbiting as planets do. 'Deterministic' seems not the right word for this sort of situation.

However, quantum theory has a technical way of describing the physical state of a particle – or of a system including many particles – in terms more precise than 'blur'. This description is in terms of a 'wave-function' or 'quantum state'. How that works mathematically needn't concern us here, but it allows one to know with great precision the probability that (if we take a measurement) we will find a particle *here* or *there*, or with *this* momentum or *that* – without implying that the particle has any definite position or momentum when we are not measuring it. There is an equation, 'Schrödinger's equation', which governs how quantum states (these precise maps of 'probability densities') change over time, and it's a deterministic equation. So far, it would seem we can hardly describe anything here as 'random', or happening by chance.

When we do take a measurement, thus linking the quantum level with our more familiar level of the universe, the 'blur' resolves itself one way or another. Only one of the coexisting alternatives as to position or momentum survives: we do find the particle in one place and not in many places at once, or with one momentum rather than many. Surely all of us are familiar enough with the way statistics works to realize that we will not *necessarily* find the particle in the place it's most probable to find it, or with the most probable momentum. Because probabilities don't allow us to say with certainty where we will find a particle or how it will be moving, it is at this juncture (with our act of measurement) that determinism and predictability, cause and effect, can be said to break down. But it's also at this juncture that probabilities and statistics kick in, in a significant way, and these are probabilities and statistics that arise from the quantum states whose changes over time, as we said above, are determined by equations governing the quantum level.

Just as statistical probabilities with regard to a vast number of voters allow an election poll expert to predict with astounding accuracy who will win an election and by what margin, without that expert knowing how any one individual will vote; so also statistical probabilities with regard to a vast number of particles govern the emergence of everyday certainties and accustomed events on the common-sense level, without our knowing where we will find any one particle or how it will be moving. The overall result is the highly predictable pattern to which Polkinghorne referred. A billiard ball or a planet is not in many places at once. A chair looks and feels like a chair. Something *completely* different really is very unlikely to happen.

The way the quantum world translates into the world we normally perceive, at that juncture where the quantum blur of alternatives reduces to one of the alternatives, is still a deep mystery. Penrose believes we'll need an entirely new theory to make sense of it, a theory that will be to our present understanding what Einstein's theories were to Newton's. Many other physicists think we are not likely ever to gain such understanding. The suggestion has been made that it is at this juncture that an opportunity arises for God to make a choice, perhaps even bring about what we would see as a miracle defying the laws of nature. Let's take a look at these possibilities.

As we'll find later in this chapter when we discuss chaos and complexity theory, in most of nature extremely small changes can have enormous consequences. It might seem that we would also find this to be true on the quantum level, but so far no-one has been able to show that it is. So, pending future discoveries in chaos theory, we must conclude that choosing that one particle will show up in one position rather than in another probably wouldn't in itself get God very far, just as determining one vote in a national election is unlikely to throw off the results the pollsters have predicted. God would have to make a choice as to the activities of a very, very great number of particles at once. If God did that, the result could be significant. If it were a result consistent with normal experience and natural laws, we might not term it miraculous or even be aware of it. Can we imagine that such manipulation goes on all the time? On the other hand, the result *could* be something extremely unusual, to the point where we would indeed dub it miraculous. Here we begin to have a problem, and it is a matter of interpretation whether this problem may be such as to rule out *any* sort of godly intervention through manipulation of the quantum level. Dawkins has provided us with some statistics which will help us understand the situation we face. The miracle he is proposing is the sort that occurs in post-biblical miracle stories, some of them contemporary, and in Ingmar Bergman films. Nothing like it is reported in the Bible.

Dawkins writes:

> If a marble statue . . . suddenly waved its hand at us we should treat it as a miracle, because all our experience and knowledge tells us that marble doesn't behave like that . . . Molecules in solid marble are continuously jostling against one another in random directions. The jostlings of the different molecules cancel one another out, so the whole hand of the statue stays still. But if, by sheer coincidence [the argument we are examining would ask us to substitute 'by God's action'], all the molecules just happened to move in the same direction at the same moment, the hand would move. If they then all reversed

direction at the same moment the hand would move back. In this way it is possible for a marble statue to wave at us. It could happen. The odds against such a coincidence are unimaginably great but they are not incalculably great. A physicist colleague has kindly calculated them for me. The number is so large that the entire age of the universe so far is too short a time to write out all the noughts! It is theoretically possible for a cow to jump over the moon with something like the same improbability.[8]

We could take the position that if an event is possible, no matter how improbable, it's not a violation of physical laws if it happens. By extension, we could allow God to do almost anything without risk of breaking laws. That 'anything' might be so unlikely as to make the distinction between improbable and impossible almost meaningless; nevertheless the distinction is there. God's activity would be blatantly interventionary but not, strictly speaking, a violation of the laws of the universe. However, it's also arguable that setting aside the statistical probabilities to that extent, indeed to *any* extent, *is* breaking laws, laws such as Schrödinger's equation which determines the 'probability densities' on the quantum level, which in turn emerge as predictable events on our level. Perhaps this is a matter of interpretation. If it is, then that leaves those who are searching for loopholes for God's intervention having to decide just where to draw the line as to which divine intervention or miracle does stretch probabilities, or stretches them too far – not an exercise we can fruitfully undertake in this book!

Instead let us look at some other situations where the physical laws, at least as far as we are able to understand them, don't strictly determine outcomes. We have already seen several examples. Some of these situations have direct quantum mechanical origins and some do not. There is the spontaneous symmetry breaking we discussed in Chapter 3 in connection with the electroweak theory – along with the analogy of the pole set on end which could fall in any direction; and the hot bar of metal which upon cooling becomes a magnet, with no way to predict from the underlying laws which would be its positive and negative poles. Or take the planets in the solar system: nothing in the laws which governed the origin of the system dictated that there should be a certain number of planets, not more nor less.

As we move up the ladder of complexity from fundamental particles to atoms, to molecules, and eventually to human beings, we find at many junctures a possibility of things going one way or another without either outcome breaking the underlying laws or being incompatible with what happens at a more fundamental level. Even in cases where, with hindsight, we can *explain* how it happened, we cannot always actually *predict* what happened, in the sense of saying this had to be the outcome, the laws would not have allowed any other. At these points it seems nature had a choice, and we are powerless to say why one choice was the result and not another.

Where science cannot tell us why things should have developed one way and not the other, the possibility is left open for us to say that we simply don't have enough information, or to suggest that no amount of information would suffice, that what happened was pure chance or that God chose the outcome. Suppose God, having set up initial laws, made later choices and fine-tuned the universe as it developed, making certain it was a universe suitable for our existence. We suspect – though without complete understanding of all the laws we can't know for certain – that the alien who has never seen our universe would not be able to look at a collection of protons, neutrons, and electrons and other elementary particles and study the laws of quantum mechanics and predict that the periodic table of elements must inevitably exist precisely as it does and could not be different in any detail whatsoever. At the same time we know that if the alien saw the periodic table of elements he would agree that it accords with elementary particles and quantum laws. Shall we suppose that it was God who without breaking any underlying laws or skewing any statistics decreed that electrons, protons, and neutrons, etc., should arrange themselves according to the table of elements as we know it, and not according to some other possible table of elements? If there were indeed choices at that juncture, that was a decision with enormous significance for the future of the universe. Shall we suppose that as human beings evolved God tipped the balance to provide them with self-awareness – another momentous choice? There is nothing we know of at more fundamental levels which determines absolutely that this prop-erty will emerge.

Though this flexibility might explain how God could have

203

influenced the over-all development of the universe without breaking its laws, it is less helpful with the question of specific events which seem to defy those laws. Even if it was not inevitable that certain properties and not others would emerge at certain points of discontinuity, we know that these properties have emerged. They do not happen only some of the time. It still looks as though laws would have to be broken for the sun to stop for Joshua or for Christ to rise from the dead.

Those who believe in God may run an additional risk with an interpretation such as this which depends upon our continued ignorance of the reasons why certain properties and not others arise. Are the arguments we've been looking at examples of 'God-of-the-Gaps theology'?

THE DEATH OF THE GOD OF THE GAPS

I have used the words 'situations where we and our science are powerless to say why things should have developed one way and not the other' as though those situations were easy to pin-point. In fact we have no clear way of recognizing such a situation. In many areas, such as the way galaxies cluster, we don't know how much we should attribute to the fundamental laws and how much to random consequences of those laws, where underlying symmetries are broken. This hampers our ability to determine what the fundamental laws actually are. Phenomena in science such as the photon don't come with a tag labelling them 'broken symmetry'. The disguise problem makes for a continuing puzzle. How much really is random about the development of the universe and what goes on in it, and how much is not?

Our uncertainty here might seem to argue that we are farther from complete understanding than we have been hoping. It also argues that we can't be sure where the gaps are through which God might be allowed to intervene.

'There is much we don't understand' is certainly a commendably modest attitude. It happens to be true. However, all statements about our present ignorance can be made only with the qualification that we have no way of knowing what we will learn in the future. The inadequacy of human understanding of the physical universe has in the past proved a risky argument for religious belief. Plugging God in wherever we have gaps in our knowledge is popularly known as God-of-the-Gaps theology. When our forebears said that there was no explanation for the evidence of design in nature except to recognize that God was the designer, they were invoking the God of the Gaps. Science, in the business of plugging gaps in human knowledge, has a way of squeezing out this God. Darwin filled in the design gap. We don't yet know who might fill in the gaps of emergent properties, even the gap of our understanding of human self-awareness. The failure of the God of the Gaps doesn't prove there isn't a God, but it does mean one would be well advised to found one's faith on something other than the hope that science won't be able to explain much of what is presently a mystery to us. Should we warn such believers that their 'gaps' may fail them at any moment?

Perhaps not? Relatively new branches of science, called chaos and complexity, deal with areas that science in the past has allowed to remain gaps – unpredictable systems. But instead of filling gaps, this science is revealing how unfillable some of them are. What we are learning is something which naive common sense always seemed to tell us, but which science was slow to recognize as enormously significant – that most systems contain elements of both predictability and unpredictability.

CHAOS MEETS CONTROL

Since the time of Isaac Newton, we've tended to believe that the universe consists of predictable systems and to assume that science is in the process of reducing everything to such systems. Some of our forebears in the eighteenth and early nineteenth centuries thought we would eventually find that everything in the universe is predictable, and Hawking and others still encourage similar hopes, with modifications to allow for quantum uncertainty and a few other problems.

However, we now find to our surprise that predictable systems are the exception, not the rule, even in areas of science which seemed most dependably predictable, such as Newtonian dynamics. A second surprise, after discovering the prevalence of chaotic, unpredictable behaviour, is to find pattern in the midst of the chaos. American physicist Joseph Ford, a leading scholar in the field of chaos, tells us that in the process of unveiling chaos we recognize that 'nature's dice are only slightly but nonetheless purposefully loaded. Our scientific task is thus to determine the loading and the purpose.'[9]

Ford defines chaos simply as randomness – a term in ordinary usage, but which needs more precise definition if we are going to employ it here.

Part of the scientific process is finding patterns in observational data. That's a human enterprise which predates science. The ability to compress experience into a mentally coded form reflecting cause and effect may have helped our ancestors survive long before language or numbers appeared. For a more up-to-date example: rather than having to describe in a lengthy narrative or show in an enormously long video how the positions of the planets in the solar system change over time, we have Newton's laws – a shorthand formula for describing and predicting these changes.

Those familiar with computers will recognize that this sort of coding is essentially what we do when we create a computer program. To cause the computer to produce the string of numbers 2, 4, 8, 16, 32, 64, 128, 256, 512, 1024, 2048, we do not type all those numbers into the computer. We can write a brief computer program to produce that string. But there are also strings of numbers which have no pattern that we can use to create such a program. Imagine flipping a coin over and over, letting heads be 'one' and tails 'zero'. Unless we are Tom Stoppard's Rosencrantz and Guildenstern – whose coin flips always came up heads, thus signalling that something had gone wrong with the universe – we will generate a random sequence of zeros and ones. The result might be:

01101000110111100010

The shortest program we could type into a computer to reproduce that precise string of numbers would be the entire list of numbers themselves. If there is no way whatsoever to represent the string of numbers in an abbreviated form – no pattern that can be used to code the information into a program shorter than the string itself – then the string of numbers is truly 'random'. The more abbreviated we can make the program, the less random the string of numbers it represents is. Ford tells us: 'For sequences, as in nature, order is the exception, chaos the norm. The number of definable patterns is countable. The number of possibilities is not.'[10] Common sense hints that there might be some pattern to be found within a random list, without that pattern being of the kind that could be exploited to code the information into a shorter program. Common sense also hints that we might program the computer in such a way as to allow some random steps in the middle of an orderly program. In fact, we saw such a program in Chapter 5 in the section on evolution.

Chaos science is the science that studies randomness: the randomness of the weather or a turbulent sea, the way smoke eddies and swirls, the way a flag snaps and waves, how liquid flows and traffic snarls, the oscillations of the heart and the brain, and the changes in wildlife populations, even the unexpected randomness in the solar system, and the way galaxies cluster. Scientists in areas which before seemed only distantly related (the heart and brain don't traditionally fall into the same area of science as the study of galaxy clustering) have discovered connections between the different kinds of irregularity and made progress in understanding unpredictability. Though there are critics who insist it's artificial to force these systems under a single umbrella, it does seem significant when we find that, in systems as diverse as sand dunes, economics, the human body, and the large-scale structure of the universe, self-organization appears to be inevitable in the midst of chaos – a trend as strong as, perhaps even stronger than, the increasing disorder (entropy) brought about by the Second Law of Thermodynamics. 'Complexity theory' is the outgrowth of chaos theory which focuses on the borderland between the predictable and the unpredictable, the delicate balance between order and chaos.

A little background is needed to understand the impact chaos and complexity studies are having and their relevance for our discussion. We are told that in Russia scientists have for some time considered the study of chaotic systems an important branch of science, and they were surprised to hear in the seventies and eighties that Western scientists thought there was anything 'new' about it. But in the West as well, the study of such systems long predated the field of study we now call chaos.

In 1961 an American meteorological scientist at MIT named Edward Lorenz was in the process of studying the weather by means of computer simulations. The story goes that one day, upon examining the printout from the previous day's simulation, he decided it would be instructive to run that same simulation over again. The computer program he was using allowed six digits after the decimal point, but he hadn't saved his starting numbers for the previous day's run. The printout was all he had, and the printout showed only the first three digits after the decimal point. When he typed the simulation back into the computer so that it could run again, he could reproduce the numbers he had used the day before only to the third decimal place, not to the sixth.

Lorenz thought this would not matter seriously. He assumed that errors so tiny, occurring only at the fourth digit after the decimal, would cancel each other out. However, when he restarted the program, he found the computer producing 'weather' that was different from the weather the same program had produced the day before. At first the changes were small, but as the simulation continued, they became larger and larger, until the weather was completely different from what it had been in the previous simulation. The tiny changes in the numbers, changes brought about by rounding off to the third decimal point, made a significant difference. Lorenz had discovered one of the principles of the mathematics of chaos: small events have enormous consequences.

Lorenz's discovery didn't come as a complete surprise to him, since he was a meteorologist. Meteorologists had long harboured suspicions that when it comes to the weather, very small causes have very large and far-reaching effects. As someone with a bent for whimsy had put it, the movement of a butterfly wing in Asia affects the weather in New York a few days or weeks later. Hence the name, 'the butterfly effect'. As it turned out, this is not mere whimsy.

The scientific name for the butterfly effect is 'sensitive dependence on initial conditions'. Tiny differences cannot be counted upon to cancel each other out as Lorenz first assumed they could. The upshot is that in any system that displays sensitive dependence on initial conditions (and such systems turn out to be the rule rather than the exception in nature) perfect predictions are not humanly possible. For instance, in weather forecasting, no matter how many weather satellites, weather balloons, barometers, and thermometers we monitor, we won't see all the details at any given moment, all the initial conditions for any system. Without that knowledge, regardless of how elegant and deterministic the equations are that we're using to make the predictions, perfect prediction is beyond us. In Ford's words, 'The source of chaos is missing information.'[11] It's the old story of garbage in and garbage out – 'garbage in' defined as data which is incomplete or incorrect in even the minutest detail.

Nearly a century before Lorenz, British scientist James Clerk Maxwell had arrived at something of the same conclusion:

> When the state of things is such that an infinitely small variation of the present state will alter only by an infinitely small quantity the state at some future time, the condition of the system . . . is said to be stable; but when an infinitely small variation in the present state may bring about a finite difference in the state of the system in a finite time, the condition of the system is said to be unstable. It is manifest that the existence of unstable conditions renders impossible the prediction of future events, if our knowledge of the present state is only approximate, and not accurate . . . It is a metaphysical doctrine that from the same antecedents follow the same consequents. No-one can gainsay this. But it is not of much use in a world like this, in which the same antecedents never again occur, and nothing ever happens twice.[12]

The eighteenth-century French mathematician Pierre Simon de la Place had theorized that an omniscient being with unlimited powers of memory and mental calculation, knowing the exact state of everything in the universe at any given moment, and knowing the laws of nature, could extrapolate from that the exact state of everything in the universe at any other given moment. However, as we've seen, one problem of trying to predict specific events (such as the Derby winner) from a Theory of Everything is precisely the problem Maxwell pointed to above: we, unlike La Place's omniscient being, can never know the exact state of things.

Another problem with prediction has to do with knowing the laws of nature. Chaos scientists find that systems which exhibit extreme dependence on initial conditions also quite understandably exhibit extreme sensitivity to the form of the equation used to figure out what a future state will be. If the equation has the smallest inaccuracy or omission, they find their prediction pointing off in one direction while the real world goes in another – with quite a dramatic degree of deviation.

All of which doesn't necessarily mean that La Place was wrong. He called for an omniscient being knowing the exact state of everything. How exact would that knowledge have to be? How far beyond Lorenz's three decimal places? Would prediction be possible for a being for whom no information is missing?

A little more background . . . after which we'll be prepared for that discussion.

When certain equations produce the same results over and over again when 'iterated' enough times, the repeating result is an 'attractor'. To 'iterate' an equation means to solve it repeatedly in sequence in the following manner.

Let us say we are simulating on a computer the fluctuation of an animal population, and we have a number representing the population for year 1. We plug that number into our equation and solve the equation, which tells us what the population will be in year 2. Knowing the population number for year 2, we use that as our start-up population figure and run the equation again to find out the population for year 3, and so forth. We're 'iterating' the equation.

In 1971 Robert May used this method to study animal populations. One of the variables which May could play with in his equation was the 'rate of population increase', which he called R. When he raised that rate to a value greater than 1, an interesting pattern emerged. As he iterated the equation, the population fluctuated for a few years and then it came to a standstill. As May continued the iteration, the population figure didn't change. The equation kept producing the same result year after year. When an iterated equation is drawn toward a number like this and remains there once it has reached it, that number is an 'attractor'.

When May raised the value of R (rate of population increase) to greater than 3, something even more bizarre happened. The population figure settled to two attractors (called 'bifurcation'), alternating yearly between the same two numbers. At a value of

R greater than 3.4 is settled to four attractors. When R (rate of population increase) was more than 3.57, there were no attractors – no pattern – 'chaos'. However, May found new organization showing up in the middle of the chaos: attractors, bifurcation – looking, when he graphed his results on a computer, like a miniature of the first attractors and bifurcation – and then later a miniature of that – and of that – and so forth. The repetitions were not exact, but the pattern was recognizable (see Figure 6.1). Mitchell Feigenbaum of the Los Alamos National Laboratory in New Mexico later discovered that the distances between the bifurcations in all bifurcation diagrams (resulting from any equations that produce bifurcation diagrams) grow smaller by a constant ratio – another mysterious element of order within chaos.

The Mandelbrot set (see the illustration section and the book jacket) is the most familiar example of the 'fractal' quality, the self-similarity on every scale, which May found in his diagram. Within the fantastic swirls of colour and design produced by Benoit Mandelbrot's mathematical scheme, we find repetition on infinite levels of magnification, but the repetitions are not exact. For more homely examples of fractals, there are cauliflowers, the way frost forms on windows, the structure of snowflakes, the branching of trees.

In addition to attractors of the sort May found in his study of animal populations, there are other kinds of attractors in chaotic systems. Lorenz had used computer graphing to study convection (the way heat moves through air). Although the convection equations continued to demonstrate the same sensitive dependence on initial states that he had discovered earlier, his graph displayed an astounding predictability – an 'attractor', but not one exactly like those we talked about above. The line on the graph described what might be seen as a twisted figure eight, and it traced it over and over again. However, the pattern never repeated itself precisely. The line never even crossed itself. The line in the 'Lorenz attractor' as we usually find it pictured seems to cross itself only because we are seeing it in two dimensions. Lorenz's graph is three-dimensional. Where it seems to cross, it is passing in front of itself.

We call what Lorenz found a 'strange attractor'. Lorenz, however, didn't call it that. He published a paper reporting what

Figure 6.1

ATTRACTORS, BIFURCATION, CHAOS, AND ORGANIZATION WITHIN THE CHAOS:

GRAPHING FLUCTUATIONS IN AN ANIMAL POPULATION:

Screen 1: Robert May found that when the rate of population increase was low, the population became extinct (left end of curving line). When it was higher, the population would reach equilibrium – settle on an attractor. When it was higher still, the population would oscillate between two alternating values. That is 'bifurcation' (where the line divides on the screen). When it was very high, the population figure would fluctuate unpredictably (chaotic region at right side of screen)

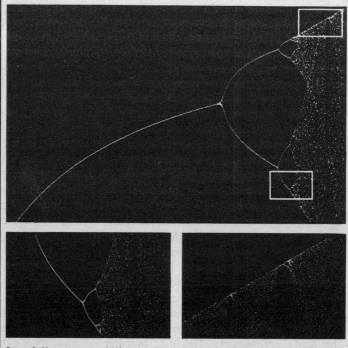

Screen 2: New attractors and bifurcation. This screen is a magnification of the lower box drawn in Screen 1

Screen 3: A magnification of the upper box in Screen 1. We find a new attractor and bifurcation, organization showing up in the midst of the chaos

he had discovered, but it was largely ignored at the time. His discovery wasn't recognized for what it was until more than a decade later. (To be strictly accurate, it hasn't yet been proved to fit the present mathematical definition of a strange attractor, though it has all the expected physical properties.) Strange attractors are found in nearly all chaotic systems. Since many scientists define chaotic systems as those containing strange attractors, that is perhaps not surprising.

With this rudimentary summary in mind, we can return to the question of how small a change will make a difference in the future of a system and whether it is possible for any being, human or divine, to predict or control future events.

Peter Coveney and Roger Highfield wrote in their book *The Arrow of Time*:

> Once an irreversible dynamical system has been sucked into a strange attractor, it is totally impossible to predict its long-term future behaviour. This is because . . . strange attractors show incredible sensitivity to the initial conditions: unless the system is started out with initial conditions of literally infinite precision, it will end up being completely unpredictable. Although the differential equations governing the way these irreversible systems evolve through time are deterministic, in the sense that knowledge of the initial conditions suffices in principle to predict the entire future behaviour, their exquisite sensitivity smashes the dream of a clockwork and predictable universe.[13]

Are Coveney and Highfield correct that our inability to know the infinite details of the initial conditions in any system really 'smashes the dream of a clockwork and predictable universe'? If human beings can't predict something, does that mean it is inherently unpredictable? If we agree that the clockwork is smashed, we might be guilty of the 'can't study it so it can't exist' attitude mentioned in Chapter 3, or of redefining 'reality' to mean only 'what we can know' or what is useful to us. Among those who support the attitude that the clockwork is smashed are Ilya Prigogine and Isobel Stengers, who in their book *Order out of Chaos* wrote: 'When faced with these unstable systems La Place's [omniscient being] is just as powerless as we.'[14]

Not everyone agrees that what we have discovered in chaos

213

theory is *inherent* unpredictability. When Coveney and Highfield say that 'knowledge of the initial conditions suffices *in principle* to predict the entire future behaviour', that 'in principle' is of no help to human beings, but it leaves a door open for omniscient beings. Ford, discussing the knowledge of the initial state required to determine orbits with Newton's equations, writes:

> Don't Newton's equations provide us with a relatively simple computational rule for determining each orbit once the initial state S_0 is given? They do indeed, but who provides the initial state S_0? . . . only a god can provide the initial state S_0. Thus, we have now tracked our missing information to the real number system whose individual members are, in general, beyond man's ability even to define, much less specify or compute.[15]

Was La Place right after all? *Could* God provide the initial state? Could an omniscient being predict everything, if that omniscient being had infinite knowledge of everything in the universe at a given moment, and knew the correct form of the equations? Of course if an omniscient being didn't know all that, that being wouldn't fit the definition of omniscient. To know the individual numbers of the real number system, every butterfly wing, every molecule of every wing, with infinite precision, isn't possible for humans. If we posit a God who does have all that information as well as the equations, and if we believe that chaos theory is on the right track, where does that get us with regard to the legality of God's intervention in the universe?

First, we have spoken of 'infinite knowledge' and 'all the information'. What degree of precision are we really talking about? According to chaos theory, however far into the future one wishes to predict the behaviour of a system, there is *some* degree of accuracy of knowledge of the initial conditions which would make it possible to predict. Complexity theory, on the other hand, has it that neither the near future nor the distant future of a chaotic system is predictable; it is the realization of a random process; hence there is no faster way to learn about the future of the system than to sit around and wait for it to happen. We must keep in mind that in almost any real-life system there is the possibility of some input from outside the system. May's animal population simulation assumed that there would not be a sudden drought or a new predator; in real life we can assume quite the opposite. While we are happily predicting future orbits in the solar system, a comet might collide with a planet, resulting in a change in the planet's orbit. We would have to expand our question of predictability to take in a larger system, including where and when the comet originated. Start anywhere and we are likely to find ourselves eventually having to take into account the entire universe, including ourselves the observers. Will even that be enough? Must we take God into account as well?

Those who believe in God welcome chaos and complexity for other reasons beyond the fact that these theories reveal gaps which human knowledge will never fill: (1) these theories can be seen to demolish the concept of a deterministic, mechanistic universe; (2) chaos theory appears to allow an omniscient being to determine events through infinitely minute changes in initial conditions. It might seem that someone is trying to have his cake and eat it too. Reason number one is cause for rejoicing among a great many of our contemporaries, regardless of religious orientation. Reason number two is more problematical. Chaos and complexity theories at present leave too many questions unanswered. Attempts to speculate about what God could or could not do using the tool of sensitive dependence on initial conditions presuppose a great deal of knowledge about God and the universe that we do not yet have, and probably never will have. Let me show you what I mean.

We will begin with the most extravagant speculation, that with the discovery of sensitive dependence upon initial conditions we have found a way God might have been able to set up initial conditions at the birth of the universe so that everything, for as long as the universe endured, would happen down to the minutest detail exactly as God had planned it from time zero. Even the miraculous events of the Bible would have been predestined from the beginning, and it would be possible to accept the words of the Midrash: 'The Almighty entered into an agreement with all that was created in the Six Days of Creation . . . that the waters should split before the children of Israel . . . the sun and moon should stand before Joshua . . . the fish should vomit Jonah . . . the fire should not harm Hannaniah, Mishael and Azariah . . .'[16] This is not a view explicitly expressed in the Bible, but if God could have contrived the initial conditions in such a way that every event, miraculous or not, in all history would happen inevitably (and through 'natural causes') because of the precise fine-tuning of those initial conditions – then we have certainly found a happy resolution to the contest between science and religion – and we can all ride off into the sunset just as God planned we would from the moment of creation!

However, though sensitive dependence on initial conditions might seem to give God a tool for that sort of setting up, there are serious problems with this proposal. In the first place, it can be argued that those deterministic initial conditions needn't have been set up by God – and we are right back to the old clockwork determinism. But for any *purposeful* setting up, God would almost certainly have to have knowledge beyond what La Place seemed to be allowing his 'omniscient' being when he spoke of complete knowledge of the exact state of everything *at one instant in time*. God's infinite knowledge would have to include infinite foreknowledge. The prevalence of randomness discovered in chaos and complexity science would seem to make this *less* likely, not more so. If blind chance (or such other possible wild cards as human choice) *ever* play a role, God, if he lacks a way of knowing in advance which way such free chance and choice will go, will have a problem. *Does* God have infinite foreknowledge, or perhaps stand outside of time so that all knowledge is present knowledge? It goes without saying that chaos and complexity theory can tell us nothing about that.

Here is a further suggestion. God's foreknowledge might come from awareness of something about the initial set-up which itself would determine which direction would be taken at every future 'random' fork in the road – a scheme in which what are made to look like throws of the dice are not really random, but somehow predetermined from the very beginning. However, much of what chaos and complexity are telling us would indicate that there *is* real contingency, that in the history of the universe the forks in the road where chance or choice played a role have been too numerous to imagine. Is there real contingency? Or are the dice so loaded as to make a mockery of randomness? Our attempts to answer those questions have only just begun.

Another question with which chaos and complexity don't help us is the question whether *any* deterministic set of initial conditions for the universe – without further input from God – suffices to bring about later behaviour so contradictory to the norm as the sun stopping for Joshua or a resurrection from the dead.

Let's turn to a suggestion which doesn't have God affecting the universe only through initial conditions at the beginning of the universe. This proposal is that God, with infinite understanding of past and future, could act in the universe at any time and bring about miracles and answers to prayer by infinitesimal, unnoticeable manipulation. Flicking a butterfly wing to cause the winds and waves to calm in the Sea of Galilee doesn't seem such a mammoth manipulation of nature after all. We must again acknowledge the problem: unless God were to work very close to hand indeed, wouldn't God need to know (or control) not only present events, but also some future events at junctures which appear to us to be random?

Ford sees God not controlling all specific details, but able to use probabilities to bring about an end result:

> To my mind, the randomness of chaos permits one to imagine a much more believable, even likeable deity than does the traditional predestination based on Newtonian dynamics. In a chaotic universe, God fits naturally into the role of riverboat gambler or Las Vegas casino owner. Having set the probabilistic rules of the game, he need have no knowledge of who specifically will or will not lose at the game of life. But even though God lets each 'gambler' win or lose strictly according to the rules of the game, it is the slight probabilistic edge the 'house' gives itself which implies that the final outcome will accord with the wishes of God. With only the outcome fixed, God can join humanity in the bleachers with no foreknowledge of how the 'local game' will turn out; he can with us expectantly watch events unfold . . . a deity might find a chaotic world worth watching. He knows the beginning and the end, but what happens in between is anybody's guess![17]

Ford's concept of God supports our conclusion in Chapter 5 that God could have set the probabilities in such a way as to know from the beginning that beings capable of responding to him would evolve.

Ford's casino God is, on first thought, appealing. The Mississippi gambler – taking risks – enjoying the game . . . until we remember that it is *you and I* this God is risking. 'Join humanity in the bleachers'? *Are* we in the bleachers? Surely not. We're down in the muddy field! This God might be, in fact, none other than the 'God with the hands-off policy' we discussed in Chapter 5 – unless we go on to suggest that if a 'player' asks for help from God he or she might receive it; that God might, upon invitation or of his own volition, jump off the bleachers into the mud and join the game, provide special equipment and coaching, or send in hot-shot visiting players. Could God – even without leaving his comfortable seat as an onlooker – predetermine in the dice some interim outcomes, short of the 'end' Ford speaks of? Perhaps later prayers might be foreknown and factored into the loading from the outset, some themselves shaping the ultimate goals that God chooses to advance. Ford's scheme even without these further embellishments, would seem to require that God must have foreknowledge, or have a great deal of past experience with other universes in order to judge how to set up the probabilities. We must remind ourselves, however, to be cautious about assuming that, if the dice are loaded, it is necessarily God who did the loading. We can imagine such loading occurring without divine purpose within the framework of any of the First Cause competitors we discussed in Chapters 4 and 5.

One great unanswered question which looms over all this speculation is the question of the role of human choice. The accounts in the Bible, as well as later history and common experience, would lead us to think that when Abraham, Lot, Moses, David, St Paul, you and I, and the rest of humanity make our entrance, the drama takes interesting twists. The characters in the Bible for the most part certainly act and talk as though they had plenty of free will and exercised it continually, often with ludicrous and disastrous results for themselves and the best-laid plans of God. On the other hand, God as described in the Bible usually knows ahead of time where this exercise of free will is going to lead, and God is ready for it.

Let us take for example the story of Jonah.[18] In order for this story to occur, we need a sea creature large and hungry enough to swallow a man whole without digesting too rapidly, improbably positioned in a precise area in the eastern Mediterranean. We need Jonah on a ship passing nearby. We need a particularly violent storm. And we need the specific throw of dice that causes the other men on board ship to conclude that it's Jonah who must be thrown overboard if the ship is to survive. The weather, the sea, the migration and feeding patterns of enormous sea creatures, the dice – can we suppose an omniscient being could set it all up without having to break any laws or any deterministic chain of cause and effect? Never mind all that – we have a far greater difficulty with Jonah himself.

Recall the story. God commanded Jonah to go to Nineveh and preach to the people there. If Jonah had done that, and the rest of the story had happened on the journey to Nineveh, we could say that this was all part of the perfect plan of God. But Jonah, apparently not an automaton but endowed with free will, chose not to obey God. Jonah was appalled at the idea of preaching to the Ninevites, and he lodged a serious complaint with God. The Ninevites were renowned rotters who would probably kill him. On the other hand his preaching might cause them to repent and be forgiven by God – and thus escape the damnation they so richly deserved! Neither outcome was acceptable to Jonah, and he refused to go.

As the story has it, God heard these arguments but nevertheless stuck to his guns, at which juncture Jonah again exercised his free will and set off in the opposite direction from Nineveh. Only because of that did God find it necessary to line up boat, fish (or whale), storm, and dice so as to deposit Jonah in Nineveh after all, get the Ninevites preached to (they did repent, just as Jonah feared they might), supply a metaphor for Christ's death and resurrection (three days in the fish, three days in the tomb), provide us material for a rousing spiritual two or three thousand years later, and give me material for this chapter. None of that would have happened if Jonah had not disobeyed God.

What are we to make of that? Regardless of what we make of that particular story (and there are reasons besides not believing in miracles for thinking it is a parable, not history), one of the most profound questions we can ask about the universe is whether we do indeed have such choices as Jonah had and you and I appear to have. If we do, what role do they play in the long-range pattern? Next to that question, all the others having to do with emergent properties, quantum theory, symmetry breaking, and other instances of unpredictability dwindle to insignificance. Not knowing whether our thoughts and actions are predestined by God, truly our own, or predetermined in a biological or mechanical way renders us powerless to judge whether or not our activities add a serious element of unpredictability. Common sense tells us that surely they must, and that a God attempting to work his will in the world would have to contend with this problem. It is interesting to note that in human behaviour we also find a balance between unpredictability and predictability. Perhaps Jonah's decisions could not be predicted precisely, but the patterns of human nature he exhibited are as familiar as those of our next-door neighbour in the twentieth century.

Let us finally return to the question of 'gaps' with which we began this entire discussion of chaos/complexity. Though it is not an answer preferred by most scientists, who would rather not find so much of the universe inherently beyond our knowing, there is nothing yet in science to forbid one's thinking that God could have intervened at many junctures as the universe evolved and could be intervening regularly today. Wherever and whenever things look random, God might step in and determine then and there how those particular dice landed – when symmetries broke one way and not another, when galaxies clustered one way and not another, when mutations appeared which allowed human beings to evolve who could respond to God, and at many other junctures in our day-to-day lives. Science insists such continual meddling is not beautiful, but that is a judgement based on the aesthetics of science, not necessarily the aesthetics of religion – or on what really happens. Chaos and complexity lead us to suspect that there are far more of these junctures than we previously assumed. In fact, our picture of the universe has shifted so drastically from the deterministic picture some of our forebears had of it that many see it as evidence of God that we have even the level of organization we do observe. Those who would prefer to leave God out of it think there must be an as yet unknown organizing principle at work, and they hope that science will eventually discover what it is.

In summary: do chaos and complexity theory allow us to reconcile the idea of an intervening God with a rational universe? In fact, these theories reveal a situation that may include such immense freedom and flexibility as to make the legalistic approach look more than a little ridiculous. Certainly an approach to the possibility of divine activity which presupposes a deterministic, mechanical universe is far too constraining. The old dichotomy between Intervening God (or Intervening Us, for that matter) and a Clockwork Universe has crumbled. The picture that is emerging is subtle and complicated in ways we are only dimly beginning to understand. In a universe which combines predictability and freedom, as these theories suggest ours does, insisting there was a violation of the fundamental laws of the universe if the waters parted for the Israelites or even if Christ rose from the dead might be tantamount to insisting that the Constitution of the United States is violated if traffic is allowed to flow the wrong way on a one-way Fifth Avenue for several hours to accommodate a St Patrick's Day parade.

Chaos and complexity are also providing a fresh way to understand an old paradox by allowing us to see that chance and choice, on the one hand, and necessity, on the other, are inherent properties of the universe, not in conflict, but working in tandem to allow the universe to be rational and patterned and at the same time contingent. After all, that does describe the common-sense world we experience and observe, and the one both the scientific method and religion assume we have.

But, when all is said and done, if we want to be even more anti-legalistic, why should God, if there is a God, be burdened by *any* 'necessity'? Is God free not merely to join in the game operating under the existing rules, but to change the rules at whim – and even to insist on a different game entirely? 'Death' in Ingmar Bergman's *The Seventh Seal* upsets the whole chessboard when the Knight seems to be winning; Death always wins, regardless. Can't God the Creator do whatever he pleases, whenever he wishes? We'll get back to these questions in due course.

'TOP-DOWN' DETERMINISM?

Our discussions of chaos and emergent properties may fuel a suspicion that reductionism (the idea that everything can be explained in terms of its most fundamental components) is not a valid scientific concept; that we have more room than we thought

221

to speculate about which level of description is more significant or 'real' – perhaps even which is most fundamental. Is it the common-sense level in which a chair is something that supports my weight and I don't worry over much whether a baby universe is being born under my left eyelid? Or is it the uncertain, phantasmagorical world of quantum mechanics? Are all the levels of description equally real? Are perhaps none of them real? The following short digression will help prepare us for a more revolutionary suggestion of how God might intervene in the universe.

Because quantum physics undermines many of our common-sense beliefs about reality, we might conclude that the common-sense level of description is an illusion, while the quantum physicist has got the right picture. Niels Bohr, one of the founding fathers of quantum mechanics, is usually associated with the view that independent reality (in the sense of such things as inherent certainty of position and momentum independent of our measurement) doesn't exist on the quantum level. His long-standing disagreement with Einstein on this point provides a lengthy chapter in the history of the philosophy of science. However, when it came to the everyday level of the universe, Bohr put a lot of stock in the common-sense description and stressed that we couldn't discuss or adopt a quantum mechanical point of view at all if we didn't first have the common-sense point of view. Whatever we know makes its first entrance through this window. Bohr wasn't merely being condescending.

Bohr would have had us believe that my common-sense view of the chair across the room from me isn't a 'mistaken' or naive view which is then refined and corrected by the quantum physical point of view. He didn't believe either view was of the chair as-it-is-in-itself.

Among other interpretations of reality on the quantum level is the 'many-worlds' view introduced by Hugh Everett that all probabilities are realized – all possible results occur. Wherever there is the possibility of something happening one way or the other, both possibilities actually do happen. For instance, I find the electron here, while an alter-ego finds it there. Carry this through with an infinite number of such possibilities, and we find alter-egos and even parallel universes proliferating at a mind-numbing rate. We get an awesome array of 'realities'! On the other hand, there is the opposite view that the role of measurement and observer is so strong that it is meaningless to speak of any other reality on the quantum level than observer-influenced reality – perhaps even observer-created reality. Yet another interpretation still holds to the belief that the uncertainty we find on the quantum level is a product of our ignorance and the inadequacy of our measuring capabilities, not of inherent uncertainty.

Since we have so far no clear understanding of how quantum reality (whatever it is or isn't) becomes common-sense reality, and in view of all we have seen above about the breakdown of predictability at so many points along the road to greater complexity, it seems not unreasonable to reject entirely the tyranny of the quantum level of the universe, either as the level which determines everything else or as the level on which we have our surest and most fundamental view of reality. Could we not speculate that it may be the *more* complex levels that determine the less complex levels, not the other way around? This is an intriguing possibility, because apparently we human beings are at present the most complex level, at least on our own cooled cinder. Is it our existence which determines what the other levels are like and what properties must emerge in each? Must everything from the particle level up exist in a way which would allow *us* to exist?

This is not as far-fetched as it seems. In fact, we have been here before, and it ought to sound rather familiar. We have come full circle to the anthropic principle, which has it that we observe the universe as we do because if it were different, we wouldn't be here to notice. We can argue on our behalf in this master-of-the-universe contest that it would be a little ridiculous to think of the fundamental particles as determining what the rest of the universe is like, or of their being more 'real' than things on our level, when, in one way of thinking about it, we and our measuring devices determine what *they* are doing.

It takes a little of the wind out of our sails to remember that if we can invoke the anthropic principle many others besides us also have (or had) a right to do so. 'We are It,' said the hyracotherium (the earliest known member of the horse family, about the size of a small dog). 'If the universe hadn't been set up precisely right, we couldn't be here. We hyracotheria observe the universe as it is

223

because if it were different we hyracotheria couldn't be here to observe it. The universe exists because we exist.'

Having just knocked ourselves off the pedestal as the inevitable ultimate goal of the universe, we can climb back up again. Because, although many levels of development have the 'right' to invoke the anthropic principle, that doesn't mean they do or did. It takes a certain level of complexity to have the ability to think of the anthropic principle.

Nevertheless – off the pedestal again! We have no reason to assume that we are the ultimate in complexity. In fact, it's fairly safe to assume we are not. If the process of evolution continues, what emergent properties will yet arise?

Suppose everything *is* best explained from the viewpoint of complexity rather than from the viewpoint of simplicity, what does that imply about the possibility of a creator? We think that the universe progresses from simplicity to complexity. That's the way it seems to have happened chronologically. There were elementary particles before there were atoms, atoms before there were molecules, and so on up the ladder. If that's true, then it makes sense also to assume that in order to create the universe, a creator would have had to be there doing something at the beginning of time. The 'First Cause' had to be there FIRST, chronologically.

Since the general chronological trend is from simple to complex, that would imply that God, if God was here first, must indeed be the simplest. You'll remember that Dawkins ruled out God on the grounds that God, as most people conceive of God, was much more complex than some of the things God is supposed to have created – such as DNA.

But if we open up the new possibility that it is the more complex levels (such as human beings or even more complex creatures yet to come) whose existence determines what the lower levels must be like and what properties must arise in them, then God might be not the simplest of all (sorry, Bernstein!), but the most complex of all – and all the levels might 'emerge' from the most complex level . . . a sort of top-down creation in which not the First Cause, but the Final Cause or end-purpose dictates and constrains all that precedes it. Since 'most complex' doesn't imply largeness (we certainly aren't the largest thing in the universe), this concept would allow us to discard the notion that we are too small for God to notice.

When we consider this scheme, we're implying something more than the anthropic principle. We're no longer merely saying that the future determines the past in a passive way. We're suggesting that something in the future actively and consciously creates the past, and we have been searching for the creator and intervenor in the wrong direction. To make this work, time as we know it would probably have to take second billing to a larger framework of time, or timelessness.

'I AM'

There is an old Texas aphorism: 'Time is how God keeps things from happening all at once.' Perhaps for God things do happen all at once, and 'time' as we know it is only an approximate description.

As long ago as the fourth and fifth centuries the Christian philosopher Augustine of Hippo gave a great deal of thought and prayer to the subject of time. Like Aristotle and Islamic natural philosophers, Augustine concluded that time begins with the beginning of the universe. He made a sharp cut between the things that exist in time and space and what is outside time and space.

Augustine began with the question 'What was God doing before He created Heaven and Earth?' and decided that the question has no meaning because words such as 'before' and 'after' and 'then' can't apply where time as we know it doesn't exist.[19] According to Augustine, time as we know it is part and parcel of this creation, not something that applies to God.

The timeless present tense in which Augustine proposed that God exists is difficult to imagine or describe. Augustine wrote: 'Who shall lay hold upon the mind of man, that it may stand and see that time with its past and future must be determined by eternity, which stands and does not pass, which has in itself no past or future.'[20] Augustine doesn't say, you will notice, that eternity lasts for ever, though that's how most of us think of eternity. Eternity lasts no time at all. Eternity 'stands and does not pass,' and 'in eternity nothing passes but all is present.'

In this model of reality, you can't talk about a 'time' before

225

time was created, any more than you can talk about it in Hawking's no-boundary universe. There was never a 'time' when time didn't exist. 'There can be no time apart from creation . . . Let them cease to talk such nonsense,' wrote Augustine.[21] What he proposed instead of 'such nonsense' was that God, existing in an eternal present, creates chronological time for the benefit of our human minds and existence.

What would it be like if events were not ordered in chronological time? If God knows everything in the universe that ever has happened and ever will happen in the same way (except in infinitely more detail) that I know what's happening right now in the room with me, in what way would that affect God's power to affect this universe? What meaning could cause and effect have in such a setting? What would happen to 'predictability'? Where events are not filed chronologically, is there some other sort of filing system? Those are questions we have no hope of answering, but we can speculate a little.

Our chronological framework forbids knowledge of the future. That's a prescription one wouldn't have in a timeless situation. It wouldn't be at all surprising to find God knowing the future – it would all be NOW to God. That makes problems for us, because it is difficult to think of ourselves as having free will if someone knows the future and knows what we are going to decide. However, I know what I did yesterday. I decided to push on with this chapter rather than to write some long-overdue letters. It would never occur to me that this knowledge, which I have on Wednesday, in any way obliged me to make that decision yesterday, on Tuesday. True, I can't change my mind about it now. Is it my knowledge about what I decided yesterday that makes it impossible for me to change that now? Why should I necessarily conclude it is that?

We cannot assume it is knowledge of the past that robs us of ability to change it. Why should we assume that knowledge of the future robs us of our ability to change the future? Why, in any instance, should knowledge of an outcome determine that outcome? In our framework of chronological time, knowing the future *would* seem to determine the future, and certainly the psychological situation of knowing and having free will at the same time would not be one we could cope with – a good argument for why that possibility isn't allowed in our spacetime

ABOVE: The Large Magellanic Cloud: the mass and distribution of visible matter in our Milky Way galaxy is not enough to account for the orbit of this satellite galaxy – a clue that a mysterious halo of invisible 'dark matter' must surround our galaxy. (*Harvard College Observatory/Science Photo Library*)

BELOW: Map of the universe 300,000 years after the Big Bang, produced from data from the Cosmic Background Explorer (COBE) satellite in spring 1992, shows the fluctuations in temperature that represent ripples of wispy matter which later may have evolved into the stars, galaxies and clusters of galaxies we know today. (*Associated Press*)

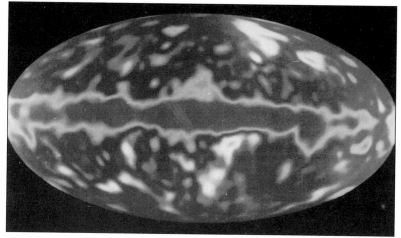

RIGHT: William Blake: 'The Ancient of Days', a God whose mathematics, symmetry and geometry circumscribe the universe. (*British Museum/The Bridgeman Art Library*)

BELOW: Michelangelo: 'The Creation of Adam', a God who created us. (*The Mansell Collection*)

ABOVE: From the Raphael Loggia in the Vatican Museum, by Perino del Vaga of Raphael's workshop: 'So the sun stood still, and the moon stopped, till the nation avenged itself on its enemies.' If the Old Testament story of Joshua is correct, God intervenes in the universe and breaks the laws as we know them. (*The Mansell Collection*)

LEFT: A terracotta figure of Christ by Giovanni della Robbia, *c.*1500. In an even more astounding intervention, according to the New Testament, God 'became flesh and dwelt among us'. (© *The Board of Trustees of the Victoria & Albert Museum*)

RIGHT:
David Hume, engraving by W.J. Edwards: 'What have we to oppose to such a cloud of witnesses [to a miracle], but the absolute impossibility or miraculous nature of the events which they relate?' (*The Mansell Collection*)

BELOW:
Thomas Aquinas: 'As the other sciences do not argue in proof of their principles, but argue from their principles to demonstrate other truths in these sciences, so this doctrine does not argue in proof of its principles, which are the articles of faith, but from them it goes on to prove something else.' (*The Mansell Collection*)

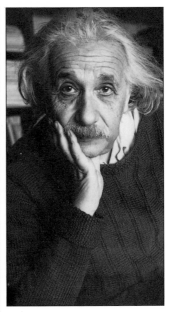

ABOVE LEFT: Sir Isaac Newton: 'It is the perfection of all God's works that they are done with the greatest simplicity... As they that would understand the train of the world must endeavour to reduce their knowledge to all possible simplicity, so it must be in seeking to understand the [Scriptural] visions.' (*The Mansell Collection*)

ABOVE RIGHT: Albert Einstein: 'God does not play at dice.' (*Popperfoto*)

LEFT: Niels Bohr: 'Stop telling God what he can do.' (*Hulton-Deutsch*)

ABOVE: In Tom Stoppard's play, Rosencrantz's and Guildenstern's coin tosses always come up heads, signalling that something has gone awry with the universe. John Stride (left) and Edward Petherbridge in the National Theatre production of *Rosencrantz and Guildenstern are Dead.* (*Photograph by Anthony Crickmay. From the collections of the Theatre Museum. By courtesy of the Board of Trustees of the Victoria & Albert Museum*)

BELOW LEFT: Chaos physicist Joseph Ford: 'In a chaotic universe, God fits naturally into the role of riverboat gambler.' (*School of Physics, Georgia Institute of Technology*)

BELOW RIGHT: Meteorological scientist Edward Lorenz, who discovered one of the principles of chaos: small events can have enormous consequences. (*The MIT Museum Collection*)

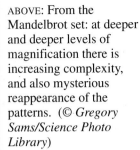

ABOVE: From the Mandelbrot set: at deeper and deeper levels of magnification there is increasing complexity, and also mysterious reappearance of the patterns. (© *Gregory Sams/Science Photo Library*)

MIDDLE: Turbulence pattern from the wing tip of an aircraft. Such turbulent systems exhibit chaos with structure on every scale, sometimes self-similar and sometimes not. (© *ONERA*)

BELOW: Frost on a windowpane, a natural fractal. (© *Pekka Parviainen/Science Photo Library*)

ABOVE: C.S. Lewis: 'What one must not do is to rule out the supernatural as the one impossible explanation.' (*Hulton-Deutsch*)

LEFT: Sir Brian Pippard, Cavendish Professor of Physics, University of Cambridge: 'Can we discover any way of combining public and private knowledge into a complete description? Only by joining to our outward-looking skills those employed by expert cultivators of the inner landscape and preferably by combining both kinds of expertise in the same individuals.' (*By courtesy of Sir Brian Pippard*)

creation. But why should this necessarily hold for God in a regime where time as we know it doesn't exist at all? It isn't difficult to imagine a situation in which I have free will and God might know every last detail of what I'm going to do for the rest of my life. As a seventeenth-century Afghan writer expressed it, 'All the pages not yet written He has read' – and yet I can write on them anything I choose.

The biblical description of God's activity in the world makes a great deal more sense if Augustine's model of time is the correct one: God's ability, as described in the Old Testament, to plan over a period of thousands of years, taking into account all the spanners that his Chosen People are going to throw into the works; the blame that falls on Judas, though Judas' betrayal of Christ fulfils prophecy; puzzling incidents in which Christ apparently overlooks the fact that his disciples are constrained by a chronological point of view and has to re-explain in a way that will make sense to them; Christ's statement 'Before Abraham was born, I am'[22]; and all the incidents of prophecy, great and small. None of it seems so bizarre if God is seeing it and intervening in the whole of 'history' at the same instant, not constraining our free will but taking advantage of our choices and mitigating the consequences. The oddness from our point of view is merely the oddness with which this perfectly feasible activity shows up in our chronological time, where it doesn't mesh and we have no vocabulary to describe it.

We, of course, have no idea whether this is the way time works – or the way God works. We do know that we can't yet understand time. It remains one of the great mysteries. We suspect that the chronological arrow of time as we know it is a broken symmetry, because the underlying laws of physics don't in general have an arrow of time themselves. With few exceptions, they are time-reversible. If a law allows a sequence of events to occur, then it also allows a time-reversed version of the same sequence – the film run backward. Nevertheless, in most of nature, events and change occur in a time-directed manner and the film is never run backward. Once again, as in the case of galaxy clusters, it's difficult to determine whether what we observe is really a broken symmetry or something more fundamental. The best judgement at present indicates that chronological time is only a part of a more fundamental reality.

More fundamental reality . . . more real . . . less real . . . Must different 'realities' compete? Some have suggested that when we speak of physical reality and spiritual reality we are dealing with things so different and separate that ne'er the twain shall or should meet: there is material reality in which the laws of nature operate dependably and unwaveringly; and there is spiritual reality in which God presumably acts. Of course the proposition we are considering in this chapter is quite different – that God is active in the physical realm as well as the spiritual realm in such ways as to make it not sensible to talk of two separate realities. Perhaps we would be better served by the concept not of two separate realities, but of two separate descriptions of the same reality.

'Complementarity' means using two different, perhaps mutually exclusive, descriptions in order to gain a better understanding than either description alone could provide. It was Bohr's way of addressing the problem in physics known as wave-particle duality.

When we experiment with the way light propagates (the way it travels), we find that it acts as though it were waves. The description of it as particles is ruled out. When we study the way light interacts with matter, we find that it acts as though it must be particles. The model which describes it as waves is ruled out. By 1920 it was clear to physicists that light could be conceived of either in terms of waves or in terms of particles, but that neither model by itself was adequate to explain the experimental data, and that this odd situation could not be resolved by saying that light is sometimes particles and sometimes waves. By the mid-twenties physicists had found that the problem applied to matter as well as to radiation. The description of the electron as a particle of matter cannot account for all the data. There are instances in which it makes no sense to describe it other than as a wave.

Bohr, Einstein, Austrian-Swiss physicist Wolfgang Pauli, and German physicist Werner Heisenberg were the most vocal in the discussion of the wave-particle duality problem. They and their colleagues debated it, along with other conceptual problems of quantum mechanics, for several years. Would one model – wave or particle – turn out to be a 'better' representation? Could we attribute a 'more fundamental reality' to one model? Were both equally 'real'? Were neither real, both only 'useful as thought devices'?

It was Bohr who had the most to say on the subject. In a letter to Einstein in 1927, he seems to have concluded that we have to live with the contradiction. He suggested that it 'is possible for us to keep swimming between the realities' as long as we don't allow our intuitive feeling that matter and radiation must be *either* wave or particle to 'lead us into temptation'.[23] Bohr had begun to work out a new way of dealing with these contradictory 'truths', a way of thinking about them in which it makes no sense to play the one off against the other or try to decide that one is correct and the other false, a way of accepting them as incompatible but both necessary.

Those who criticize the use of similar thinking when there seems to be a conflict between religion and science point out that the two dilemmas are not similiar. With wave-particle duality, the two descriptions are mutually exclusive because they work only in mutually exclusive experimental situations. It would be difficult to show that the same is true with the descriptions of science and religion. Furthermore, with wave-particle duality everyone is in agreement about what the data is, and also about the fact that the two incompatible descriptions are both necessary. In a science–religion controversy we do not start with any such agreement. Many scientists refuse to assign any validity whatsoever to religious data which seems incompatible with scientific data. However, for those who are certain that there is an active, intervening God and also certain that science is reliable, and for whom this seems to create unresolvable contradictions, the problems *are* similar, if not completely parallel, and many find complementarity helpful.

Science and religion are, for them, two different descriptions which together give us a fuller understanding than either description alone could provide. If the descriptions are mutually exclusive, that is disturbing, but the fact that it is disturbing should, they believe, not lead us to opt prematurely for one description over the other or to find an artificial reconciliation. Like Bohr, we can accept both descriptions as less than adequate descriptions which satisfy our human need to have a description, while remembering that all visualizations and language descriptions fail us in situations beyond the power of human mentality. Others who believe in God deem this approach unsatisfactory, saying that God, in their experience, insists on occupying a front-line position in all descriptions, all conceptual schemes, all experimental situations. This God is a presence, not merely a way of thinking about or describing the universe, and not acting exclusively in realms beyond the power of human mentality.

Does complementarity help us resolve the dispute between Creationist and scientist? C. S. Lewis has suggested that even the conflicting views of creation might be explained on the grounds that neither alone is a full explanation of data consisting not only of the physical universe but also of the human mind, soul, and psychology, our moral situation, the problem of evil, and our relationship to God. Genesis and modern science are incompatible descriptions, but, as with Bohr's 'realities', the question as to which is correct may be naive, based on our intuitive feeling that it must be either one or the other. The question can be answered only in a way that would imply 'Both are correct' *and* 'Neither is correct.' Together, says Lewis, they are more nearly 'complete' than either is alone. Yet both are no more than feeble human perceptions – or revelations of God dimly understood, or revelations that God purposely made simple enough for human mentality – of occurrences which are inestimably far beyond our capacity to understand or describe. If we knew what really happened at what we call 'the beginning', we'd find our current disagreement more than slightly embarrassing, regardless of which side of the dispute we're on.

We cannot leave this subject without expressing reservations about applying 'lessons learned from physics', such as complementarity, to other areas as though their validity in physics automatically proclaims them as principles which must hold for a wider range of experience. However, Bohr himself perceived duality and complementarity in many areas of life, in and out of science. He didn't invent the idea, although he had to do some interpreting and adjusting to make it work for waves and particles. John Hedley Brooke, in his book *Science and Religion: Some Historical Perspectives*,[24] traces some of the sources which probably contributed to Bohr's thinking. Bohr's father, a physiologist at the University of Copenhagen, argued that a mechanical explanation of living organisms did not render a second explanation in terms of their meaning and purpose superfluous. If one wanted to know everything there is to know about animal behaviour, he insisted, one needed both explanations. According to Brooke, the psychology of William James, the theology of Kierkegaard, and the philosophies of Kant and H. Hoffding also probably influenced Bohr. We might conclude that, in adopting complementarity, science was forced to borrow a rather unscientific concept to rationalize a situation which seemed hopelessly unscientific. Now the same concept is lauded as a 'scientific' way of approaching other areas of knowledge and reconciling science and religion. This sort of boot-strapping is not highly recommended.

Perhaps it is instructive and encouraging to note that Dirac found a theory which succeeded in combining wave and particle in a description without contradiction or paradox. Beyond the word and picture description, there is in the maths that underlies it a flexibility which does reflect the dualism we've been discussing, but a simple mathematical transformation is all that's required to rewrite the equations of motion (that have to do with particles) as a wave equation. This did not keep Einstein from saying, somewhat later, 'All these fifty years of pondering have not brought me any closer to answering the question, What are light quanta? Nowadays every Tom, Dick and Harry thinks he knows it, but he is mistaken.'[25]

There is another approach which moves the question of incompatibility between science and religion peremptorily out of range of human discussion.

THE ULTIMATE SELF-CONFIRMING HYPOTHESIS

In one way of thinking about it, arguably the only sensible way: if God is God he can break any laws he's made at any time, owes us no apologies, and cannot be judged by any criteria other than his own. Though we can deduce what God's criteria probably are, we don't really know. An eminent nineteenth-century physicist, Sir George Stokes, said: 'Admit the existence of a God, of a personal

God, and the possibility of the miraculous follows at once. If the laws of nature are carried on in accordance with his will, he who willed them may will their suspension.'[26]

Stokes had found a gentle way of saying what the Bible states much more bluntly. 'Who is this that darkens my counsel with words without knowledge?' God asks Job in response to Job's well-justified queries and complaints.[27] And a little later: 'Will the one who contends with the Almighty correct Him? . . . Would you discredit my justice? Would you condemn me to justify yourself?'[28] Nor is this a strictly Old Testament view of God. In his letter to the Romans Paul quotes God as saying, 'But who are you, O man, to talk back to God? Shall what is formed say to him who formed it, Why did you make me like this?'[29]

This God is – let us put it frankly – the ultimate self-confirming hypothesis. We can despise such a concept of God if we choose, but we could also make a good argument that no God can be God without being at the same time the ultimate standard and definer of such concepts as justice, goodness, faithfulness, even self-consistency. Arguing from our own standards, or what we think ought to be God's standards, has dubious validity. We grant ourselves authority, as Evelyn Waugh has one of his characters say in *Brideshead Revisited*, to 'set up a rival good to God's'.[30]

Shall we do that? Whether or not there is a God to rival, shall we set ourselves up as the ultimate judge of what is good, just, faithful, and self-consistent? We know we may have invented those concepts, or they may have arisen because they just happened to give our ancestors a survival advantage in the environment in which they found themselves, while another environment would have encouraged completely different standards.

We might argue differently that goodness, justice, faithfulness, and self-consistency are standards that are intrinsic to the universe quite apart from ourselves and also apart from evolution or any God. The eighteenth-century philosopher Gottfried Leibniz treated standards of beauty, wisdom, and good as though they were independent of God's will or choice. In this way of thinking, God is obliged to conform to standards which are stronger than himself – a view not far from that expressed in this chapter by those who insist God is forbidden to set aside the laws of the universe because that wouldn't be faithful and self-consistent, and God must be faithful and self-consistent. But why should God necessarily be faithful and self-consistent, and who or what defines 'faithful' and 'self-consistent' anyway? Whence came these principles and definitions? Are *they* somehow the self-evident, self-confirming hypothesis and therefore more powerful than God?

Perhaps we have overstressed the point, but it isn't so obvious or easy to accept. We have examined several candidates for First Cause: a mathematical and logical consistency which makes the universe inevitable, the universe itself (via the no-boundary proposal), ourselves (via the anthropic principle), and God. We have assumed that it would be possible to support a choice among these First Cause candidates on some other basis than self-confirmation! Yet, while perhaps despising the notion of God as the ultimate self-confirming hypothesis, we find shockingly upon reflection that it isn't only God who presents us with the problem. Wherever the buck does stop, wherever we find the uncaused First Cause which has no explanation or reason for being – be it mathematical and logical consistency, no-boundary universe, human beings, or God – we have discovered a self-confirming hypothesis. For no reason is it *thus* . . . for no cause . . . it simply is. Ultimate and complete truth is beyond proof or meaningful argument, partly because at that level there is no point of view outside it from which to judge, no standards external to it by which it can be tested, none not defined and set by itself. Such truth is by its very nature unprovable, unfalsifiable, and self-confirmed. 'I AM' – full stop.

One longs to insist this is not so, that God or a theory of absolutely everything (no matter how compelling its logical self-consistency) is never merely self-confirming because of the requirement that it must show its consistency with reality as we observe it. However, there are sticklers who would argue that if a First Cause claimant can validate its claim only by linking itself to observational evidence, then it isn't a good First Cause candidate at all. Why? Because then observational evidence – reality as we so inadequately observe it – becomes a stronger standard and concept than the First Cause claimant being tested against it. The bottom line of this argument is that if there are *any* independent standards against which a First Cause candidate can be judged, then that candidate isn't First Cause.

This becomes a hopelessly esoteric discussion. It is a little ridiculous at this interim to worry too much whether we might discover, on the brink of ultimate knowledge, a true conflict between logical self-consistency and observed reality and perhaps moral standards as well. After all, the same sticklers mentioned above would point out that if we get to be the judges as to which is the strongest concept, and are competent to say there is a conflict, that makes *us* the ultimate source of all standards!

Leaving this maze of fruitless speculation aside for the moment, let us simply concede that if the 'I AM' *is* God, the creator and final standard of everything, we don't have to like that God, and if he breaks his laws or seems to break his laws, we aren't allowed to say that he can't. The same Bible which claims we are created in the image of God also quotes God saying: 'My thoughts are not your thoughts!'[31] When the Pharisees criticized Jesus for appearing as his own witness, he told them: 'Even if I testify on my own behalf, my testimony is valid.'[32] As we've seen, the end of the Book of Job makes the same point with implacable severity.

Nevertheless, God as described in the Bible, though not unwilling to remind human beings of his absolute power, *does* set himself standards of goodness and faithfulness, and vouchsafes to humans at least partial knowledge of what these standards are. When God's activities appear to be at odds with what we have been led to believe are his standards or with what we have learned through science, it would seem there ought to be a way of dealing with the problem other than throwing up our hands and exclaiming that God is a self-confirming hypothesis.

THE MASTERFUL USE OF PARALLEL
PERFECT FIFTHS

A model of reality which would be more palatable to most of us and still in keeping with both the scientific and the biblical viewpoint would be to think that the physical laws of the universe, and standards such as goodness and justice as we are aware of them, are only a part of a larger picture.

When we talk about the rational universe, what we really mean is the universe as we observe it and what we have been able to extrapolate from those observations – the universe as we have managed to make sense of it. Chapter 3 should have shaken your faith in that picture just a little. We have heard Barrow suggest that the reason why the universe is intelligible to us is that we have singled out areas which make sense and ignored others. Chaos scientist Joseph Ford has written: 'If analytically derivable as well as chaotic orbits are unpredictable, one begins to suspect that Newtonian dynamics is but a theorist's fanciful description of a perfect world inhabited by perfect observers.'[33] These are not 'anti-science' statements or arguments that the picture of the universe we get from science and scientific theory is subjective and artificial, having nothing whatsoever to do with what really is 'out there'. They are merely a recognition of the point-of-view problem we discussed in Chapter 3. Recognizing the pervasiveness of that problem puts us in a far stronger position in any quest for bed-rock truth.

Saying that we don't have the whole picture in view isn't the same as saying that what we do have in view is a false picture. I can discover much that is true about my own village without knowing anything at all about the countryside surrounding it or the cities, oceans, and other continents that lie still farther away. Even though some would argue that without seeing my village in a larger context there is much I won't understand about it, and that I'd certainly better not assume that knowing my village gives me great expertise about the rest of the world, I still would insist that what I've learned about my village is not an illusion.

In the same way, we needn't think that if there is a more fundamental reality than the one encountered in our science, it necessarily follows that science is giving us a false picture. Jumping to such a conclusion is not the way human knowledge progresses. Even confining ourselves to discussion of physical reality, we find such theories as inflation theory which suggests that, were we able to understand our entire observable universe, that would no more be 'complete knowledge of reality' than my knowledge of my village is complete knowledge about the world.

Such an outlook is not foreign to science. Our assumption of unity, that there are underlying natural laws which do not change over time and are valid everywhere in the universe, makes it possible for us to study the cosmos and its history, even back to eras which we can never hope to observe directly. In regimes where Newton's laws fail us, Einstein's more fundamental explanation does not, but Einstein did not show that Newton was 'wrong'. When relativity led us to the conclusion that the universe began in a singularity at which all these laws would be meaningless, we took a step back, regarded the matter from the point of view of another theory, and posited such things as imaginary time, which would allow us to see our common-sense time as an approximation – but not 'wrong'. It is common practice in science to deal with logical problems, the apparent breakdown of laws, and seeming contradictions and anomalies by assuming that if we understood a more fundamental scheme of things there would be a law that would not break down, no contradictions or anomalies, and no logical problems.

We can hardly be accused of taking an intellectually untenable path, then, if we allow ourselves to speculate that all of reality and rationality as we may eventually understand them through science and common sense may not be the most fundamental underlying level of reality and rationality. We needn't at the same time relegate what we do know to the level of illusion, a mock-up film set. It may be a superb approximation. We also needn't assume that if we could find the ultimate level of understanding, we would find God there. But if there is a God, we don't have to devalue what we know and what we have accomplished in order to concede that God might know much more and operate in ways inconceivable to us.

If there is a God, and God does intervene in the universe in ways such as the Bible describes, how can we explain the fact that science, superbly powerful tool that it is, is not able to discover the deeper reality in which an intervening God would be consistent with natural physical laws? Clearly we don't know the answer to that, but there are several suggestions.

First, maybe science's rejection of all evidence that cannot be confirmed publicly, though a practical necessity, does create a severely restricted point of view. Second, maybe such things as the three dimensions of space and one of time, the forces (be they four or one), matter, and energy – perhaps our entire physical universe and the laws that govern it – are set up as a cosy environment for human beings and human minds. This is an environment which leaves plenty of room for our intelligence to exercise itself but does not present us with things we are not equipped to handle. If T. S. Eliot was correct that 'Humankind cannot bear very much reality', God or evolution has perhaps fashioned a sort of nursery situation for us in which we are shielded at least temporarily from what human mentality hasn't the capacity to comprehend and bear.

Third, perhaps human reason and knowledge will eventually evolve to a level able to understand the deeper levels, by the same process by which we continue to discover deeper laws of physics. We just aren't there yet.

Fourth, perhaps God himself will in due time – in this universe or the next – reveal what we can't find out for ourselves, and St Paul was correct to say: 'Now we see but a poor reflection as in a mirror; then we shall see face to face. Now I know in part; then I shall know fully, even as I am fully known.'[34]

However that may be, we can speculate that the complete legal code, the laws which really are never broken or changed, not just of our universe but of everything, includes a far higher order and rationality than what we presently see as physical laws. The elegance and beauty, the goodness and justice of it, may put our present standards to shame. In such a model, the laws we know would not be wrong, but they would be approximate laws, valid in certain regimes but not all, consistent with more basic underlying laws, but giving us by no means the full picture of the possibilities inherent in those underlying laws. What seems to us supernatural and arbitrary might appear perfectly natural and rational. What seem to us to be inviolable standards of justice, goodness, and faithfulness would also be shown to be approximations of a deeper, underlying standard. This hypothesis is, of course, unfalsifiable. It is, however, a hypothesis which accords well with the way 'laws' work not only in science but also in areas of human creativity.

Here is an analogy which comes from my musical training. In elementary music theory classes, students learn to harmonize chorale melodies according to rules drawn from the compositions of Johann Sebastian Bach. Bach himself didn't set down these rules; musicians and musicologists have studied Bach's music and derived these rules from what they found. The rules are very good rules, and if you follow them meticulously you can end up sounding almost like Bach! One of the rules we were taught is 'Never write parallel perfect fifths.'

Since this is not a book about music theory, I'll not explain what that means. Suffice it to say, the rule isn't merely something invented to cause grief for music theory students. In my day, if you, the student, should happen to slip up and write some parallel (or 'consecutive') perfect fifths in a chorale harmonization, the instructor was likely to embarrass you by playing your composition on the piano. Without being warned what to listen for, everyone in the class, including yourself, was jarred and annoyed when the parallel perfect fifths were played. It sounded wrong. It definitely wasn't Bach. Even someone with no technical musical training would probably have recognized that something was amiss. The 'law' against the use of parallel perfect fifths is one that reflects a certain reality having to do with what our ears find pleasing in a Bach-type chorale, and what they don't.

It is very surprising to find, then, that the chorales Johann Sebastian Bach wrote himself contain quite a few parallel perfect fifths. The rule forbidding them is a rule which Bach himself violated without the least compunction; in fact it's doubtful whether he was even aware such a rule might exist. In any case, Bach knew when breaking the rule was more correct than obeying it. It never sounds wrong when Bach does it.

Whatever the true rule is which determines when parallel perfect fifths are OK in a Bach chorale harmonization, it hasn't been discerned by those who write music theory texts – or at least it hadn't yet been figured out in the 1960s when I was in music school. It requires genius beyond what is normally available to most of us to know when the rule we know can – perhaps we might even say *must* – be broken. It requires genius which probably needs no rules whatsoever, but in whose creation rules are found to apply.

It isn't so iconoclastic to suggest that it might require genius beyond even that available in the Royal Society or the entire company of Nobel laureates to know when the laws of nature as we know them can, even *must* be set aside, in order to assure consistency on a deeper level. The Wagner fans among us will know what I mean when I say that if there is a God the Meistersingers of science had better be prepared to allow God his Prize Song.

We do not know that the single, unvarying truth behind everything is God, but the other side of that coin is, let's face it, that we also do not know what the laws are which really cannot be broken!

WHO IS THE 'I' IN 'I AM'?

The intellectual milieu in which the scientific method arose was a celebration of the potential of human beings and the human intellect, an attitude you and I adopt when we speak of setting out on a quest for ultimate knowledge about the universe. We are also well in the tradition when we claim a right, not just as 'humanity' but as individuals, to seek knowledge and understanding, and a right to choose what we will personally accept as truth.

The intellectual spirit of the times today does not favour a full range of choices. For instance, it allows us to maintain the belief that science – or if not science as we know it, some greater manifestation of human reason – will eventually comprehend everything. It is also acceptable to take the opposite view that this is 'hubris', and to envision what is beyond our reason as very grand – far more so than we – such as the Mind of God as dimly perceived in the laws of the universe, Einstein's God, or a deeper, far more fundamental set of unchanging laws. Though there might be much disagreement and debate, such views are not an outright embarrassment at high table in the colleges of Oxford or Cambridge. A God who might insult our intelligence is another matter.

Brian Pippard writes that the scientist 'is right to despise dogmas that imply a God whose grandeur does not match up to the grandeur of the universe he knows.'[35] Even if we don't believe in God, we like to think that if it should turn out there is a God this God would be our sort of God, one who lives up to our standards of rationality, justice, love, and grandeur. Where does that leave the elderly lady with blue hair and pink curlers who sits weeping while a choir sings soppy hymns on the TV screen? Have we any grounds but elitism for arguing that God should not be as accessible to her as God might be to myself or Brian Pippard, although the Jesus she thinks she knows may or may not match up to the grandeur of the universe we think we know?

The intellectual spirit of the times also places humans above any system of values. Is this carrying Enlightenment thinking to an absurd extreme? True, we say we must uphold human values, but we decide what they are. Haven't we always set ourselves up like that? That is, after all, what the story of Adam and Eve is all about. But we have traditionally relinquished some of that right by deciding that there are standards which overrule our human judgement. At least we have paid lip-service to such external standards. Today, for all we may talk about the impossibility of proving which of several choices is really the First Cause, are there any serious contenders vying with ME for the position of ultimate self-confirming hypothesis?

Unfortunately, or perhaps fortunately, our choice or right to despise or espouse dogmas or opt personally for one First Cause over another, or even create a God to our liking, doesn't actually prevent independent reality from being exactly whatever it is. Perhaps it is not 'tasteful'. If I should by any chance find myself in heaven, will I shudder at the musical style? Will it be 'contemporary Christian music' or rock rather than Bach? Now THERE is an ultimate question that *really* matters to me! Nevertheless, I don't think I have a choice in the repertoire of the heavenly choir or that, if God exists, God necessarily adheres to my tastes. Ultimate reality doesn't have to suit *anybody* – not Sir Brian Pippard or Stephen Hawking or myself or Billy Graham or the elderly lady by her television. It doesn't even necessarily have to make the least sense to us, in spite of our assumption that it will.

At the beginning of this book, I said that the only thing of which I can be certain is my own existence. (Even that is an assumption, but one I chose to make.) Why all this fuss then about humans thinking they know the answers or what the answers ought to be? Who else *could* possibly decide what is truth? I am, not by claim, but by default, my own ultimate authority on everything else.

That is the strongest possible statement of the power of the individual and of human reason, and it really is rather lame. Being by default *my own* ultimate authority doesn't guarantee I've got it right. It doesn't make me THE ultimate authority. One of the assumptions of science and religion, that there is such a thing as objective truth, means that I might be dead wrong. Of what possible worth, then, in this quest, is my private view of the universe?

7

INADMISSIBLE EVIDENCE

I should not believe such a story were it told me by Cato!
ANCIENT ROMAN ADAGE

PAUL DAVIES WROTE, IN HIS BOOK *GOD AND THE NEW PHYSICS*, 'THE true believer must stand by his faith whatever the evidence against it.'[1] In the light of the preceding chapters, we might conclude that evidence against it isn't so easy to find in modern science. But let us suppose Davies had written instead: 'The true believer must stand by his faith whatever the lack of evidence *for* it.'

We do not read of the biblical patriarchs having that level of faith, nor the twelve apostles, nor the saints of the Middle Ages. Their belief, according to descriptions in the Bible and medieval literature, was supported by direct evidence gleaned from their own personal experience. Such experience might convince you or me, *if* it actually *were* our experience.

In Chapter 5 we imagined an alien who has never seen our universe asking: 'Can you show me any reason to believe there is a God, except for the fact that science can't prove there *isn't* a God?' Where is the evidence?

241

PUBLIC VS. PRIVATE KNOWLEDGE

The traditional view has it that a significant difference between science and religion is that science is 'public knowledge' while religion is 'private knowledge'.

As we saw in Chapter 3, one of the principles which underlies the scientific method is that the testing, the direct experience of the universe, must be public, repeatable – in the public domain. If the results are derived only once, if the experience is that of only one person and isn't available to other objective observers who attempt the same test or observation under approximately the same conditions, science must reject the findings as invalid – not necessarily false, but useless.

We've seen that there are difficulties which prevent our entirely living up to this ideal of science. We find ourselves questioning whether we ever do have direct experience of the universe, whether our observations – even those which seem most straightforward and most widely corroborated – aren't always to some extent bounded and directed by our expectations and our point of view rather than by external reality. Not all evidence is as public as we would prefer it to be, because of the enormous expense of major physics experiments and the fact that astronomical observations and biological data such as the fossil record are often unrepeatable.

Nevertheless, the general principle stands: scientific knowledge is public knowledge and is tested and honed in a public arena.

Can we say the same for religious knowledge? The conventional answer is no. We can expect that when believers present their case the evidence they offer will include information and insights coming from individuals. Judaism and Christianity lay heavy emphasis on the Bible (Judaism, of course, only on what Christians call the 'Old' Testament), a collection of books pieced together from ancient manuscripts, some of which in turn were selective transcriptions of oral history. The Bible itself relies in large part on the testimony of individual people, of prophets, and of a man who claimed to be God incarnate. There are often no corroborating witnesses to biblical or other religious experience, and when there are they can't be called disinterested and objective. Religious evidence may be unrepeatable evidence, appearing at a particular time and place, with a significance that escapes any but those directly involved. There are no sure-fire formulas; there is no possibility of specific prediction from previous religious experience which would allow us to test such experience in a rigorous manner. Results derived once from prayer and obedience to God are not necessarily repeated identically for the next person who prays the same prayer and achieves the same level of obedience. Furthermore, religious evidence may be evidence that is available and meaningful only with a prior commitment of belief, but is unavailable or meaningless to a detached observer.

Again, this is over-generalization. Though many miracles recorded in medieval and Counter-Reformation miracle books, and those reported today, seem to take place for a private purpose with no public significance, the biblical account is of encounters with God and miracles which almost invariably have significance reaching far beyond the private event. The criticism that evidence is meaningless without a prior commitment of belief is a double standard we human beings readily apply to anything we personally find unconvincing. Some aesthetes and religious fundamentalists apply it to science, and scientists even apply it to one another when they suspect that a leap of faith caused by an attractive theory has provided too strong a set of spectacles-behind-the-eyes. Regarding the subjective nature of religious knowledge, there are those who argue that religion has a harder, more objective edge to it than science, because the revealed reality of God can't be coaxed or forced out by testing, manipulated or scrutinized at will, or co-opted for our use. Polkinghorne writes: 'Neither prayer nor blasphemy is a magical lever which can be used to act upon God to make him demonstrate his existence.'[2] Believers insist that to encounter God is to encounter something unmistakably independent of ourselves, to an extent that scientific evidence can never be.

There is also an extremely important sense in which religious experience is not exclusively – or even perhaps mainly – private knowledge. It has been accumulating over a far longer period of history and from a far wider sample of the population than that from which scientific knowledge has been amassed. Nor is religion necessarily private knowledge in that it is only selectively available. One of the basic tenets of Judaism and Christianity is that knowledge of God is available to anyone who really bothers to look for it, regardless of class, educational background, intellect, taste, virtue, or age. In fact, it's partly this supposed universal accessibility of God that makes religion distasteful to an intellectual elite. To find that the experience of God reported by a rocket scientist or a saint may to a significant degree be consistent with the experience of God reported by an illiterate farm labourer, a punk rocker, or Miss America is jarring to our sense of the appropriateness of things, but if it's public knowledge we're after – here it is on a scale science is hard-put to match!

Yet religion remains essentially private knowledge if we consider where the most significant testing of religious evidence takes place. Who is it that ends up convinced, or not convinced, by the evidence? However we might try to answer that question, from history or from contemporary religion, whatever some might insist the answer *ought* to be – who *should* have the last word – and regardless of the way some of us try to influence one another, we can't avoid the conclusion that human decisions about whether to believe there is a God – and, if so, what sort of God – *are* private decisions, not a consensus.

This may seem an unorthodox conclusion, running counter to formal church authority, but it isn't. Judaism and Christianity both recognize that at their profoundest level they are about relationships between human beings and God, and they provide plenty of precedent for thinking that a decision whether or not to establish such a relationship is never really a group decision. Jacob must wrestle alone with God, in spite of all his family connections. Job must stand alone before God, not along with his friends. One 'lost lamb' must be brought back, one 'prodigal son'. All heaven holds its breath, as though the fate of the universe hangs in the balance, while one man or one woman chooses whether or not to believe; but not for the outcome of a debate between British academics as to whether science has erased the need for God. Doctrine, dogma, argument, discovery, any process by which religious knowledge is externalized or authorized, any means by which pressure is brought to bear, are significant primarily as they serve as a means to that end, establishing a human–God relationship. Augustine was referring to this shift of focus away from intellectual assent to the human–God relationship when he ended one of his discussions about time and eternity with the words 'Need it concern me if some people cannot understand this? Let them ask what it means, and be glad to ask: but they may content themselves with the question alone. For it is better for them to find You [God] and leave the question unanswered than to find the answer without finding You.'[3]

There are nevertheless many of us who insist that, however private that decision might have to be, we need to see some public evidence in order to make it. Some even insist that the sort of evidence which is admissible to science is the *only* sort of evidence they allow in their private court-room. Is that so?

ADMISSIBLE EVIDENCE?

When individual human beings try to determine what is true and what isn't, they don't confine themselves to the same methods and tools they use when making such decisions in groups. There is not one among us, no matter how dedicated to the scientific method, who insists on believing only what has been winnowed out as 'truth' by that method. Some spout that rhetoric, but no-one lives by it. All of us accept plenty of evidence coming to us by means having nothing to do with a scientific method. For most of us, a class of evidence which we treat as very admissible indeed is the evidence of our own first-hand experience.

In the process of arriving at my one-woman consensus about a chair, the universe, or God, suppose I discover that something in my private experience is contradictory to public experience. Will I allow public experience to overrule me? If no-one has ever seen fairies at the bottom of my garden, and I see fairies there among the zinnias, what then? I might suggest that the contradiction is between my private experience and public *expectations* based not on 'truth' but on *previous* experience – the implication being that I'm right while public expectations are wrong. I can argue quite correctly that this is an approach I learned from science, which doesn't necessarily assume that private experience is wrong. In fact, recognizing the possibility that private experience is sometimes correct while predictions and expectations based on previous public experience are in error, and the further possibility that subsequent testing might support the new experience and lead everyone to alter previous conclusions and expectations, is one of the prods by which science moves itself along.

There is, alas, another possibility. When there is a serious inconsistency between private experience and public experience, one should doubt one's perceptions, admit the possibility of hallucination, perhaps in an extreme case even question one's sanity. I may be crazy as a loon. It is safe to assume that not every pious churchgoer or raving lunatic who believes he or she has 'encountered God' has actually done so.

But clearly, when it is *my own* experience which is at odds with public experience, the decision as to who is correct, I or 'they', isn't likely to be weighted in favour of 'they'. When we are sure of our experience, even though we may not understand it, not all the physicists, church authorities, or contrary evidence in the world can cause us to decide that we haven't experienced what we're certain we have experienced.

If this is the way human beings operate, and certainly it seems to be, then a God bent on convincing us to believe in him would be well advised to utilize a private channel. Direct personal experience would be a shortcut. Though science might have to declare such private evidence inadmissible, a human being on a search for truth does not. We must also face the unsettling thought that such direct experiential evidence, if it exists, would shed rather a different light on whatever public knowledge has been gathered.

THE SPECTACLES-BEHIND-THE-EYES, REVISITED

We have come to the brink of what may be an insuperable divide. If it's possible for a human being to have personal experience of God, then those who have had that experience, and those who have not, may have little to say to one another, not even enough for a meaningful argument. They'll see evidence in a different light, just as someone who has witnessed the murder looks at all other evidence in the court-room in a different light from the judge and jury, who were not witnesses but are trying to decide objectively from the evidence who the murderer is. One is reminded of a line in the play *The Night of January 16* by Ayn Rand. The woman accused of murder, confronting another direct witness who is contradicting her story, spits out the words 'One of us is lying, and both of us know which one it is!'[4] At first hearing, that seems a meaningful statement. It is, of course, a true statement, but it helps the audience, and the jury chosen from the audience, not at all.

We saw earlier the criticism that religious evidence is meaningful only after a prior commitment of faith, and therefore isn't admissible evidence. If I have direct experience of God, I will admit that religious evidence will be more meaningful to me, but I will also argue that my 'prior commitment' is itself based on very strong evidence indeed. I may not automatically accept all other religious evidence uncritically, but my outlook regarding everything from the Bible to science to the faith of the elderly lady with blue hair will be decidedly different from someone without first-hand experience. I may engage in what Polkinghorne calls 'rational inquiry into what our experience leads us to believe is actually the case',[5] an undertaking approved by science, but I'll do so starting from a different base of experience.

If there is no such thing as first-hand experience of God, then this faith based on first-hand experience is clearly faith based on delusion, not evidence, a faith which leads me to evaluate the other evidence in a false light. 'Rational inquiry' would be 'rationalization' of false preconceptions.

THE CLOUD OF WITNESSES

We have spent more than one chapter finding out all the reasons why we can never be sure we're observing 'independent reality' in science. If the situation is that ambiguous in science, it's surely much more so in religion. How could we possibly take seriously anyone's claim to know that he or she has experienced the reality of God?

Let us recall an argument we heard coming from scientists in Chapter 2: there must be something 'real' about what they are discovering, because otherwise it would not fit together in such amazing and surprising ways, and also because otherwise researchers would not so often find themselves confronted with the unexpected. We have seen other possible explanations – the argument that our evolutionary history has formed us so that we're able to find pattern, cause and effect, and 'fitting together' even when they don't exist; the argument that the unexpected arises only because this filtering system of our brains and consciousness isn't perfect; and the argument that we find 'fitting together' because we have made an unconscious habit of focusing on problems which are likely to have that sort of solution.

When we hear religious people use similar words: 'There must be something "real" about what we're experiencing, because otherwise it would not fit together in such amazing and surprising ways, and also because if we were making it up we would not so often find ourselves confronted with the unexpected,' we can of course similarly offer a wealth of alternative suggestions. However, we've not let such quibbles stop us from taking a lot of science seriously in this book, and it would be intellectually dishonest to apply a double standard.

Therefore, before proceeding further, we shall give the floor temporarily to those who claim they have experiential evidence of God, and allow them to clarify what they mean by such evidence. Those who are uncomfortable with this sort of testimony may skip over the next few paragraphs if they wish, but hearing these people out is in the spirit of this book, and some knowledge of what their evidence *is* will help us later in this chapter.

When it comes to the way religious belief should be applied in this world and among human beings, there is a mind-boggling lack of consensus among believers. However, when it comes to the nature of experience of the *presence* of God, there is an astounding degree of consensus. The following statements, in order to keep us as close to the source as possible, come not from the past but from our contemporaries, from persons with whom I have spoken directly. They are, however, echoed throughout the history and literature of religion.

The experience is usually not 'spooky'. It sometimes, though definitely not always, might be termed 'mystical'. It doesn't for the most part consist of events which by their nature overturn or challenge the laws of science. (I've heard only one first-hand account of an event which, if it really happened, would be very difficult to explain by *any* process presently known to science.)

The experience doesn't establish a hot-line to God, by which all questions are answered, all doubts set aside, and complete understanding is reached. It does not often provide startling new knowledge or insights to be revealed to the rest of humanity. Persons are quick to point out that, though they think their experience really is of God, it is, even at its clearest and best, only a partial, human, inadequate view of what God really is and what God is really doing. Experiential evidence sometimes comes in a flash, but it's more often the accumulation of more subtle experiences over a period of time.

John S. Spong, Episcopal Bishop of the area of New Jersey in which I live, is not by anyone's definition a fundamentalist. In fact, his rejection of orthodox biblical interpretation shocks even some who don't believe in God. Spong has said of his own experience: 'I do not mean to suggest that I have arrived at some mystical plateau where my search has ended, where doubts are no more, or that I now possess some unearthly peace of mind. Nothing could be further from the truth. I have only arrived at a point where the search has a validity because I have tasted the reality of this presence, if ever so slightly.'[6]

As to finding God initially, some say they came rather gradually to a realization that the God they'd learned about in books, songs, and from other people, is real, knowable to them personally. Others on the contrary battered the gates of heaven (when heaven was only hypothetical to them) with very sceptical demands for answers, IF such a heaven existed. Their uncompromising intellectuality led them to try to pin God to the wall in ways that might be expected to elicit a lightning bolt rather than a blessing. Their requirements for evidence and proofs were seldom met exactly as specified, but there was a moment in the process when they realized to their astonishment that they were wrestling with a real being who couldn't be contained in human descriptions or standards, not a concept or an abstraction. This God was something out of their control, something not fashioned in the image they had formed in their mind, not something that can be 'got on one's side' and used to one's advantage, 'not a tame lion', as C. S. Lewis put it in his *Chronicles of Narnia*. Regardless of how weak or strong their hopes or doubts had been, or even how great they may have thought their faith was previously, this realization was a blockbuster.

According to these reports, whatever the first encounter was, the relationship subsequently proved to be demanding, rewarding, and also sometimes disturbing, beyond prior expectations. The testimony is of God's leadership being requested and received at turning points where human foresight and knowledge were inadequate, and of God's leadership turning out to be absolutely on target, though perhaps not in the direction one would have preferred. God has led some to take risks they would never have dared take on their own and to set and achieve goals impossibly high. God has brought some through difficulty and overwhelming adversity in unexpected ways; has taken some through hell itself and not left them there; has removed all fear, in circumstances when there was every reason to be afraid. God has stopped some persons dead, when they did not want to be stopped, on the brink of serious mistakes. God has changed some in ways human beings can't change themselves even with all the help of psychotherapy. God has made it possible for them to love the unlovable, forgive the unforgivable. God has also forgiven *them* the unforgivable, allowing them to forgive themselves.

Has this all been 'spiritual' help? Not according to these witnesses. God is a powerful and active God, intervening wherever, whenever, and through whatever avenue he pleases. The phrase 'the insidiousness of God' comes from a woman Episcopal priest. God's intervention is not always kind, gentle, or pleasurable. He refuses to play by human rules or indulge our desire to be able to plan ahead. Sometimes it seems God truly enjoys cliff-hangers. God does not always come at our calling, give us what we want, or even shield us from terrible pain and grief. God sets higher standards than human beings do, and God's mercy is indeed a severe mercy, but God's forgiveness and love know no limits whatsoever.

Some direct quotes: 'My relationship with God has been by far and away the most demanding relationship in my life.' 'The Lord has been my strongest support, but also my most frustrating opponent.' 'If I didn't absolutely know this is the only game in town, I'd sure as hell get out of it!' 'The best evidence isn't some "wonder" or "miracle", and it certainly isn't success, happiness, or the peace of having my prayers answered in ways which suit me. It's the extraordinary, topsy-turvy, *interesting* course my life

250

has taken since I've engaged in this – once-begun, virtually inescapable – dialogue with God.'

There we have it. Anyone who has not had such experience can simply ignore this testimony, dismiss it out of hand, or be left struggling with the question of whether or not to believe it. Though it is widespread enough to seem hardly deserving of the label 'private', there is no way to submit it to scientific testing.

Is there *no* way for those without such experience to evaluate this evidence? Those who insist that there is a God argue that the only truly irrefutable evidence of God's existence *is* personal experience. Therefore, the only way for anyone really to find out whether there is a God, is to take at face value the words 'If with all your heart you truly seek me, you shall surely find me',[7] and call God's bluff by making a supreme, all-stops-out search, not for evidence, but for God.

Not wishing to be accused of the ultimate cultural imperialism, we will not offer that suggestion to our hypothetical alien. However, we ought to be aware that, in failing to do so, we are arguably missing the point entirely. Refusing to take the course which religion (God himself?) recommends may be the equivalent of refusing to use the scientific method to do science! However, let us proceed.

A GAME OF 'I DOUBT IT'

If we haven't had personal experience of God, any religious evidence we encounter is, at best, second-hand evidence. Even someone with personal experience would have to come to decisions about much evidence which isn't his or her own.

This is not a problem confined to religion. Most of our knowledge about the world doesn't come from personal experience but rather through reports of the experience of other people, contemporary and historical. Science has rigorous procedures for sifting through such reports in the pursuit of truth about the physical universe. Other disciplines go about their tasks using somewhat different methods from the one employed by science, not because they choose to be less 'objective' but because the evidence in their areas cannot be most effectively sifted by using the tools of the scientific method, and attempts to confine themselves to those tools alone lead to distortion.

For instance, sorting through and evaluating uncorroborated and conflicting private reports is common procedure in a court of law. Although some of the evidence heard in a court-room would not be admissible in science, it is tested in a rigorous procedure, a procedure not unlike the one we each use instinctively when trying to decide whether what we're hearing or reading is true or false.

Is the knowledge arrived at in a court of law or in individual decisions necessarily weaker knowledge than the knowledge arrived at in science? 'Arrived at' is the phrase to which we should direct our attention. Courts of law don't have the luxury science enjoys of open-endedness. Most of our legal systems require that judges and juries, short of declaring the available evidence insufficient, must muddle through and arrive at decisions. Likewise, we individuals are frequently forced to fish or cut bait. Some of these decisions are irrevocable in a way a scientific decision hardly ever is. As we've seen, it's a misinterpretation of science to think that it is in the business of rendering final verdicts.

Religion and other disciplines, partly in a defensive attempt to be more 'scientific', have become much more tentative about verdicts than they used to be. As individuals searching for truth (as opposed merely to coping with daily circumstances), we have a choice as to how tentative we're willing to be. But whenever we choose to admit private evidence from experience other than our own – and all of us, no matter how scientific by nature, do make that choice – then, either in a conscious process or instinctively, we base our decisions on two criteria: the reliability of the witness and the likelihood of the story.

'Consider the source' is a major rule of the game in *any* search for truth, even in science – which is why we attach value to our own private experience and to public experience coming from people and institutions we personally think are trustworthy. Perhaps the next strongest evidence after personal experience, when it comes to a decision whether or not to believe in God, is the evidence of those well-known to us whose quality of mind and integrity make it unreasonable to distrust their word. But even when the evidence comes from the most sterling of sources, we have difficulty believing anything which veers dramatically from what previous experience deems 'a likely story'.

For three approaches to the evaluation of reports of persons other than ourselves, we'll look to twentieth-century Oxford don C. S. Lewis, eighteenth-century Scottish philosopher David Hume, and twentieth-century Cambridge physicist Sir Brian Pippard. Lewis was among those who do not rule out the possibility of experience of God. He believed in God and was an eloquent apologist for Christianity. Hume was among those who rule out all possibility of miracles or experience of God. Many regard him as a strong advocate for atheism. Pippard, with whose philosophical approach to furniture and physics we began this book, doesn't believe in God, and yet his attitude toward the possibility of experience of God is closer to Lewis than to Hume.

THE LUCY PROBLEM

C. S. Lewis discussed the process of evaluating private evidence in several of his books and in his letters, but he expressed his own outlook nowhere more succinctly and simply than in his children's classic *The Lion, the Witch, and the Wardrobe*, the first book of *The Chronicles of Narnia*. Near the beginning of the story, Lucy, the youngest of four children visiting a large, eccentric country house, finds her way through the back of a wardrobe into another world – a snowy wood in a country called Narnia. When Lucy returns and tells her brothers and sister about her experience, they don't believe a word. The matter is eventually brought to the attention of the elderly professor who is their host. He listens carefully to their story and questions them regarding their opinion of Lucy's reliability and sanity. Then:

'Logic!' said the Professor half to himself. 'Why don't they teach logic at these schools? There are only three possibilities. Either your sister is telling lies, or she is mad, or she is telling the truth. You know she doesn't tell lies and it is obvious that she is not mad. For the moment then and unless any further evidence turns up, we must assume that she is telling the truth.'[8]

The children are dumbfounded, but the professor has closed his mind to nothing whatsoever (excepting perhaps that one should ever proceed in an illogical fashion). For instance, to pick up the conversation a little later:

> 'But how could it be true, Sir?' said Peter.
> 'Why do you say that?' asked the Professor.
> 'Well, for one thing,' said Peter, 'if it was real why doesn't everyone find this country every time they go to the wardrobe? I mean there was nothing there when we looked; even Lucy didn't pretend there was.'

Peter, of course, was asking our question: If this is real, why isn't it public, repeatable evidence?

> 'What has that to do with it?' said the Professor.
> 'Well, Sir, if things are real, they're there all the time.'
> 'Are they?' said the Professor . . .

The discussion finally draws to a close with the following:

> 'But do you really mean, Sir,' said Peter, 'that there could be other worlds – all over the place, just round the corner – like that?'
> 'Nothing is more probable,' said the Professor, taking off his spectacles and beginning to polish them, while he muttered to himself, 'I wonder what they *do* teach them at these schools.'

'I SHOULD NOT BELIEVE SUCH A STORY WERE IT TOLD ME BY CATO!'

David Hume would not have agreed with Lewis or his professor regarding Lucy's story. It was Hume who quoted the adage above, which the ancient Romans used to refer to stories so incredible they would invalidate even such an authority as Marcus Porcius Cato (Cato the Younger), an eminent statesman and philosopher in first-century Rome.

Hume was so bold as to write: 'I flatter myself, that I have discovered an argument which, if just, will, with the wise and learned, be an everlasting check to all kinds of superstitious delusion, and consequently, will be useful as long as the world endures. For so long, I presume, will the accounts of miracles and prodigies be found in all history, sacred and profane.'[9] Hume was actually referring to more than miracles and prodigies; he was referring to any supernatural claims of religion, and by extension, as he saw it, to the validity of religion in general.

In opposition to philosophers who had elevated reason over experience as a guide to truth (we will examine some of their ideas a little later), Hume insisted that all knowledge comes from experience – not necessarily first-hand experience but an accumulation of human experience. However, he pointed out that, though experience sometimes shows us that a certain effect follows a certain cause, experience has its limitations, because it can't usually show us precisely the connection between the cause and the effect or tell us how unyielding the connection is, nor does it always indicate precisely what will happen in a given situation. In some instances, experience tells us clearly that we should expect A to be followed by B, but in other instances its lesson is more ambiguous. For example, experience tells us to expect better weather in England in June than in January. But it doesn't lead us with any certainty to predict that on 1 June there will be better weather than on 1 January. What experience really indicates in this instance is that there is a probability for better weather at some times of year, but not certainty. The point Hume was making was that in drawing conclusions based on experience, there are many degrees of assurance – or, if you like, many degrees of doubt – ranging from very strong assurance that if I fall off a building I will move downward, not upward – to the sort of assurance I might feel planning an outdoor wedding in June. If we are wise, cautioned Hume, we proportion our belief accordingly, and in evaluating the truth of any reported event (a miracle, let us say) we balance the probabilities, based on experience. So far, Hume's loosening of the tight bond between cause and effect sounds as though it might allow for miracles, not rule them out.

Hume, as we've said, didn't confine experience to mean direct personal experience. He was willing to admit the evidence of human witnesses. In fact, doing so seemed to him a necessity. However, it was not a necessity that we should believe everything

255

we read or hear. Suppose the witnesses contradict one another; suppose they're of doubtful character or few in number; suppose they have a personal interest in the matter to which they're bearing witness; suppose they're hesitant and vague, or on the other hand too aggressively assertive. All of these particulars and others tend to reduce a witness's credibility. Suppose on the other hand that the witnesses are of sterling character and high credibility, but the event they claim to have witnessed is one we've never observed, which we've very seldom observed, or which has been very seldom observed in the entire course of human history.

What shall we say when faced with a Lucy, whose integrity and good sense are beyond question? Perhaps Lucy's siblings *had* studied Hume in school, but instinct would have led them to the same conclusion. Had she reported seeing someone leading a camel over the common, with no pictures or corroborating witnesses to back up her claim, the children would probably have believed her. Camels on the common were unusual, but not impossible, and Lucy could be trusted. What Lucy was reporting about the wardrobe was nothing short of a miracle. The others had no precedent for it in their own experience or in the range of human experience with which they were familiar.

Hume defined a miracle as a violation of the laws of nature as established by 'firm and unalterable experience'. We have already said that many reports of miraculous events are *not* reports of events which are in and of themselves violations of the laws of nature. In many cases it's only the claim that their occurrence is the result of direct action by God that makes us class them as miracles. If I claim I was healed of cancer because my friends and I prayed I would be, someone could argue that spontaneous healings of cancer are known to occur in cases where nobody prayed, and coincidences (such as prayer and spontaneous healing happening in close proximity) are common occurrences. There has been no obvious violation of laws of nature or those established by firm and unalterable experience. Hume's definition excludes such events as this healing from the category of 'miracle'. By anybody's definition it would be difficult to establish on scientific grounds that it was or was not a miracle.

Hume's argument had to do instead with violations which almost anybody would call violations – the regrowing of an amputated human limb, for instance, or a resurrection from the dead. He asserted that a miracle is by nature an event that has *never* been observed in any age or country. He went on to say that there is 'uniform experience against every miraculous event, otherwise the event would [not be called a miracle].'[10]

Most philosophers and logicians, even those who agree with Hume's point of view, would take him severely to task for this line of argument. Whether a miraculous event has ever been observed was, of course, the question he was trying to answer, so it was jumping the gun to *define* 'miracle' as something that has never been observed and on that ground conclude that miracles don't happen! He left himself open for Polkinghorne's remark, 'Hume turns out to be an absolutist in the matter, an intransigent sceptic who would *never* accept any evidence contradicting his prior expectation. There is no arguing with such an entrenched position, but its adoption is the antithesis of being open to the truth. It is certainly uncongenial to the habits of thought of a scientist.'[11]

We nevertheless will for the moment allow Hume the point that most of us do define miracles as events which run counter to normal and expected human experience and hear him out in his recommendations for dealing with Lucy's. According to Hume it's abundantly clear that

> no testimony is sufficient to establish a miracle, unless the testimony be of such a kind, that its falsehood would be more miraculous than the fact which it endeavours to establish . . . When anyone tells me that he saw a dead man restored to life, I immediately consider with myself whether it be more probable that this person should either deceive or be deceived, or that the fact which he relates should really have happened. I weigh the one miracle against the other, and according to the superiority which I discover, I pronounce my decision, and always reject the greater miracle. If the falsehood of his testimony would be more miraculous than the event which he relates, then, and not till then, can he pretend to command my belief or opinion.[12]

That was the situation Lucy's siblings found themselves in with regard to the story of the wardrobe, and Hume is suggesting the criteria we mentioned earlier: the reliability of the witness, the likelihood of the story. Which was more difficult to believe, that there was a country behind the wardrobe or that Lucy was deluded or lying? One would have to be Lucy herself or know her very well indeed to decide the case in her favour, and even then one could not rule out the possibility of hallucination.

Hume went on to argue that no miracle has ever been witnessed by enough people who were sensible, sophisticated, and educated enough to be unsusceptible of deception and superstition, and also of such unquestionable integrity, stability, and immunity to all the pitfalls of human nature, as to rule out their ever knowingly or unknowingly deceiving themselves or anyone else. Therefore Hume thought it would always make far more sense to disbelieve the witness than to believe the miracle.

In answer to the argument that there have been plenty of reports of miracles, so many that it isn't quite accurate to say that a miracle is at odds with public experience, Hume related three miracle stories in which the witnesses were numerous, educated, arguably objective (even opposed to the idea of miracles), and of known integrity. Nevertheless he summed up thus:

> Where shall we find such a number of circumstances, agreeing to the corroboration of one fact? And what have we to oppose to such a cloud of witnesses, but the absolute impossibility or miraculous nature of the events which they relate? And this surely, in the eyes of all reasonable people, will alone be regarded as a sufficient refutation.[13]

Polkinghorne's assessment was of course on target, and Hume's argument (quite apart from the question of whether or not we like his conclusion) was a weak and self-serving one. Better to have left matters with the weighing of the witness's reliability against the likelihood of the story. For Hume the evidence against the miracle would *always* win, because there is so much less human experience weighing in on the side of the miracle. Hume believed there was none at all, and certainly never enough to warrant establishing a religion or espousing one that is already established.

Strictly in the interest of being contrary, one might ask how many witnesses of that sort it *would* take? Hume has said there would never be enough. There is a story told of Einstein that when a newspaper article announced: 'One hundred scientists prove Einstein wrong', his reply was: 'It would only have taken one.' Likewise with a miracle. It would only take one witness. In the light of our previous discussion, we can easily predict who that one must be. If any witness could have convinced David Hume, it would have had to be David Hume.

'THE INVINCIBLE IGNORANCE OF SCIENCE'

In 1988 Brian Pippard, delivering the annual Eddington Memorial Lecture at Cambridge, asked whether human beings might ever discover a way to combine public and private knowledge into a complete description.

When Pippard spoke in his lecture of private knowledge, he didn't limit that term to knowledge which supports religion or miracles. He is by profession a scientist, but by avocation he is a musician, and private knowledge by his definition includes such things as what the Eroica Symphony means to him, as well as the colour of the chair as it appears to his eyes, and all the rest of his private mind's-eye view of the universe – all the frescoes inside his head which he can never put on public exhibition.

Pippard doubts if we'll ever know the cause of all things or whether there is a God, but he thinks we 'will not even begin to know without calling to our aid every means by which we may attain knowledge'. He goes on to say:

> It is tempting for the scientist, with the assurance he commands in his own realm, to dismiss religious experience as a delusion. To be sure, he has a right to parade the evidence that makes him sceptical of antiquated cosmologies such as religions are apt to carry in their train; and he is right to despise dogmas that imply a God whose grandeur does not match up to the grandeur of the universe he knows. But when we have chased out the mountebanks there remain the saints and others of transparent integrity whose confident belief is not to be dismissed simply because it is inconvenient and unshared. We may lack the gift of belief ourselves, just as we may be tone-deaf; but

it is becoming in us to envy those whose lives are radiant with a truth which is no less true for being incommunicable. As scientists we have a craftsman's part to play in the City of God; we cannot receive the freedom of that city until we have learnt to respect the freedom of every citizen.[14]

With the preceding discussion as our preparation, let us at last place some further evidence on the table.

'FOR THE BIBLE TELLS ME SO' – THE EVIDENCE OF SCRIPTURE

One way Jewish and Christian believers look at the Bible is as an account of human beings' experience of God during a period of testing spanning many centuries. Human ideas about God and their expectations of the ways in which he interrelates with humans were honed in the laboratory of that experience. Believers insist that the biblical account of this testing is of immeasurable value to anyone today who chooses to undertake a similar quest for understanding. They do not agree about the extent to which the Bible is literal history, or even about what it means to 'take the Bible literally'. They don't agree about how much of it is 'inspired' writing, or how much 'inspired reading' is required to benefit from it. They do not agree about all the books that should be included in it. But few believers deny that the Bible should be taken seriously.

When the Lucys who are our sources of information and evidence are far removed from us in time and culture, as the men and women of the Bible are, we are more than ever handicapped when it comes to judging their reliability. How do we know whether to trust Mary Magdalene, St Paul, Isaiah, King David, or Moses? Adding to our problem: how do we know that the report of their experiences as we find them in our Bibles is anything like the report we would have heard had we met them in person?

The Bible as it comes to us has been pieced together after the fact by editors who presumably had their personal biases. But let us suppose these accounts *have* come down to us intact as they were originally written or spoken. Let's suppose we are able to read them in the original language. We still have to ask whether we twentieth-century humans attach to these words, these turns of phrase, these modes of argument, interpretations, and literary forms, the same meaning, weight, and significance that their writers would have attached to them. How can we know how much overstatement, or for that matter how much understatement, would have been taken for granted by listeners and readers in ancient times? How do we know when biblical writers intended to have their words interpreted literally and when figuratively, when they were speaking in parables and when not? The distinction between 'material' and 'immaterial' was, according to many experts, not one that was made by primitive peoples. How does *that* cultural difference affect our reading of what they wrote and believed? How do we know which advice and rules of conduct were given for a specific situation and which were meant to be taken more generally? How do we deal with the fact that early historians used different methods for researching and writing history from those of modern historians? How do we judge the accuracy of oral accounts passed down for many generations, when we do not know what the powers of memory and total recall might have been in cultures which depended upon oral history as we no longer do?

The problem of how much of the Bible to accept as literal history – and how our acceptance or non-acceptance need affect our faith – is certainly not new. In his book *The Unauthorized Version*, Oxford historian Robin Lane Fox tells us:

> In his youth, St Augustine was turned away from Christianity, in part because of the contradictions between Luke's and Matthew's family trees for Jesus: he moved to a system of belief which took the stories of Creation as texts which could not be true literally, but had deeper hidden meanings. Yet he returned to the Christian faith and in later life wrote a massive work on Genesis, upholding its truth 'to the letter'.[15]

The question how far we can accept the Bible as accurate historical evidence continues to be debated by responsible scholars, as well as by ideologues of both extremes, but very little gets settled. The situation is so labyrinthine that we find, among scholars who don't believe in God, some who are more inclined to support the Bible's historicity than their colleagues who do believe. Fox wrote in the preface to his book, 'I write as an atheist, but there are Christian and Jewish scholars whose versions would be far more radical than mine. They will find this historian's view conservative, even old-fashioned, but there are times when atheists are loyal friends of the truth.'[16]

We live in a 'post-modern' era of literary deconstructionism, applied to all language and all written and oral accounts. Deconstructionism leads us to doubt whether words have any 'obvious' meaning, indeed any meaning at all; whether any communication is possible, whether any interpretation has more validity than another. Deconstruction has become our modern bias – and so we are also free to deconstruct *it* as an interpretation. Nevertheless, it is still radical in some divinity schools and seminaries to insist that the people who wrote the Bible must have *thought* they were making themselves clear about events, and also must have *thought* they were talking about truths which had been revealed to them! No-one denies that there was *some* definite history which took place in the ancient Middle East. It isn't quite like the quantum level of the universe! But sometimes it seems, to all intents and purposes, that it might as well be.

Again, where we stand in the great division we spoke of earlier makes an enormous difference as to how significant we think biblical evidence is. If one has personally experienced an intervening God, why doubt accounts in the Bible of similar activity? Likewise, if one has personally experienced the presence of God, why be surprised if there are inconsistencies in the Bible among stories of those who, like one's self, are faced with experiences beyond the power of human understanding and verbalization? If one's personal experience causes one to believe that God is a major actor in this world in the way the Bible makes God out to be, then attempts to treat the Bible solely in terms of secular history – automatically dismissing any intepretation in it which smacks of divine intervention – would seem to be a distortion of the worst kind, comparable to studying the Napoleonic Wars while discounting any input from Napoleon. Furthermore, arguments that no historical document comes down to us over two thousand years in pristine form have nothing directly to say to the argument that we don't really know how much input from God has affected the Bible's preservation over those centuries. Yet we must reiterate that most Christian and Jewish believers – even among those who insist they have experiential evidence of God – do not accept complete biblical inerrancy. What they do accept is that the Bible is *the* major source of evidence about God and the ways in which God relates to his creation. On what do they base this conclusion? Their answer often is that what they read there is so astoundingly consistent with their own experience of God, and that the truth of the Bible is supported by results of belief in individual lives.

IS THERE PROOF IN THE PUDDING? THE EVIDENCE OF RESULTS

One of the reasons why we trust science is the obvious results it has produced in the way of technology, medicine, and our understanding of the cosmos. Perhaps there is similar reason for trusting religion.

John Spong, whom we've quoted earlier, is probably one of those to whom Fox was referring when he spoke of Christian and Jewish biblical scholars whose interpretations of the Bible and its value as history would be far more radical and sceptical than his own, atheist, view. Yet Spong has said: 'Christianity rests its case on the evidence of the astounding transforming power we see manifested in human lives, power that makes it possible for individuals not only to change the world around them but, even more surprising, to change themselves in ways we have no right to expect humans to be able to change themselves.'[17]

As private evidence, the result of believing is strong evidence either for or against belief. If I believe in God, and am continually disappointed in what I expect of God, then I'm unlikely to continue believing. If, on the other hand, I get results such as those in the believers' statements earlier in this chapter, I will probably take that as strong evidence for God. Even lacking experiential evidence for the presence of God, I might notice that belief in God makes me stronger, better, kinder, more loving, happier, wiser, more discerning, better able to cope with life. I couldn't make a scientific experiment out of it by setting up a clone of myself to see how much weaker, worse, less kind, less loving, sadder, more foolish, less discerning, less able to cope with life, that clone would be without belief in God. I couldn't prove that it wasn't merely false belief and encouragement that was doing the trick (a sort of placebo effect), not the fact that the belief was correct. I also wouldn't be able to say how much any of the positive attributes might have *contributed* to and caused my faith rather than resulted from it. However, I might not allow those quibbles to stop my concluding that belief in God has led to good results for me and therefore must have some validity.

When we study the results in the lives of other people besides ourselves, the situation becomes even more ambiguous. What sort of evidence is it when we see a deeply religious person triumph in the face of overwhelming adversity, while his neighbour, apparently equally devout, commits suicide when faced with a similarly dire situation? What sort of evidence is it when Hawking, an avowed agnostic, also triumphs in the face of overwhelming adversity? Perhaps one knows a believer in God whose kindness, discerning wisdom, endless generosity, abundant good humour, tact, and endlessly forgiving and loving nature cause her or him to be dearly loved and trusted by all. Such a person seems strong evidence on behalf of religion . . . until one meets another believer whose fanaticism, judgemental nature, hypocrisy, unlovingness, and enjoyment of minding everyone's business bring the world nothing but grief. Again, we aren't looking at a scientific experiment. We know that many factors make us what we are, some of them deeply hidden from close relatives and long-time acquaintances, even hidden from ourselves. But are the Archbishop Tutus, Mother Teresas, and Carrie Tenbooms of this world strong evidence in support of the validity of religion? Are those whose beliefs lead them to commit atrocities and wage religious wars strong evidence against?

We have the same problem on the cultural level. There are those (not all of them religious) who argue that Judaism and Christianity have done no less than make Western civilization possible. They have been the conscience and the inspiration of Western humankind. They are the source of our values and our morality. They preserved knowledge through eras when it would otherwise have been lost, provided major themes for our art, music, and literature, and gave us the world-view from which science was able to emerge. Today many religious denominations are in the vanguard of those seeking an end to war and hatred based on race or ethnicity.

But no-one would claim that everything about Western culture and civilization has been, and is, good – and, if it is, to know the extent to which this is due to the influence of religion. We can easily argue the contrary, that much related to religion has been unquestionably evil – the Inquisition, mindless bigotry, countless sectarian wars, witch-hunts, man's inhumanity to man committed in the name of God.

Historians do not give us a clear case for judging either way. They raise our consciousness to the harm that has been done in God's name, but they also remind us that religion is often not *itself* to blame when religious arguments are used as propaganda to camouflage political manoeuvring and human activities which either have little to do with religion or are a severe distortion of religion, or when religious arguments are co-opted to uphold entrenched positions which are not really one wit more 'religious' than the ideas which threaten those positions. Historians such as England's John Hedley Brooke encourage us to revise the popular assessment of Galileo's run-in with the Roman Catholic Church. Galileo himself apparently thought he was upholding both science and religion. According to Brooke, 'To understand the predicament of Galileo in his relations with the Roman Catholic Church, it is not enough to say that science was in conflict with religion. The political ramifications of the Counter-Reformation were such that Galileo's science (which was not self-evidently correct) acquired meanings and implications that it might otherwise not have carried.'[18] Both religion and science were in the Galileo instance arguably as much, if not more, the pawns and victims of power politics than they were the cause of the conflict. And Brooke has also written: 'much of the conflict ostensibly between science and religion turns out to have been between new science and the sanctified science of the previous generation.'[19]

Surely we are unwise to rule either way on the 'evidence of results' on a cultural scale without considering carefully the political, sociological, and economic factors involved. We will look at this same topic in a more radical light a little later.

ARMCHAIR TRUTH: THE ARGUMENT FROM REASON

At the opposite pole from a reliance on experiential evidence would be a belief that faith can be validated by means of reason alone without recourse to any experience. This approach is out of fashion today even in religious circles, and in fact it has been out of fashion for nearly three hundred years. Its rejection stems partly from misunderstanding.

The arguments from reason which we're about to review were never intended to be presented to an unbeliever as 'proof' that there is a God. What Thomas Aquinas and Anselm of Canterbury undertook to do in the following discourses is what Polkinghorne spoke of as 'rational inquiry' into beliefs already held for other reasons, or what scoffers might prefer to speak of as 'rationalization' of beliefs held for no good reason at all.

St Anselm lived in the eleventh and early twelfth centuries. He was born in Italy, became an abbot of a monastery in Normandy, and later was Archbishop of Canterbury. He was a believing Christian who saw it as his task to show that his beliefs were rational, and he insisted that it was in the power of reason to show not only that the basic articles of faith of Christianity are true but that Christian belief is consistent within itself.

Anselm declared God to be 'something than which nothing greater can be conceived'.[20] In other words, whatever we may be thinking of as 'God', if we can conceive of something greater, this 'God' we've been thinking of previously isn't really God. Following that line of thought our 'First Cause candidate list' would become even more an eternal chicken-and-egg story than we decided it was in Chapter 4. If mathematical consistency is more powerful than God (if God has no other choice but to conform to mathematical consistency), then God isn't really God. Mathematical consistency is God – unless we can conceive of something greater than mathematical consistency, and of course we can. We can say 'Yes, but who determined mathematical consistency?' You can follow this line of thought further if you like. It is rather fruitless. We might even convolute Anselm's argument to prove there is no God, or at least to show that we haven't discovered God yet, because we can *always* conceive of something greater than any 'First Cause' anyone proposes.

However, Anselm's argument (his 'first ontological proof' in the classification of those who divide his argument in two) went like this. It is one thing for us to conceive of something in our minds, and it is another thing for us to understand that it actually exists. 'That than which a greater cannot be conceived' cannot be only in our minds, for that would certainly make it inferior to what it would be if it actually existed, and we CAN conceive of God's actually existing. If God is only conceived of as existing, God is not as great as if he actually exists. Since we can conceive of God's existing, he must exist or not be 'that than which nothing greater can be conceived'.

The thirteenth-century Dominican friar, teacher, and philosopher Thomas Aquinas offered 'causal arguments' for God. Aquinas didn't invent the idea. Arguments leading to the conclusion that the universe must have a cause, and that that cause is God, go back well before Aquinas and St Anselm. They are at least as old as Plato and Aristotle, and they were also advanced by medieval Jewish philosophers such as Moses Maimonides and Isaac Albalag.

The 'causal argument' has been a theme running throughout this book, and we have seen that it doesn't inevitably lead to the conclusion that the First Cause is God. There is no way to prove by science that God either is or is not the First Cause of the universe. We've discussed the suggestion that the universe itself might be the First Cause, so today even the assumption that the universe must have a cause outside itself is called into question.

Aquinas elaborated on the causal argument, and presented five arguments for the existence of God.

1. The argument from motion. Nothing moves unless it is moved by something else, but this chain of 'mover' and 'moved' can't go on to infinity. 'Therefore it is necessary to arrive at a first mover, moved by no other; and this everyone understands to be God.'[21]

As we have seen, there is reason to think this chain does not lead so inevitably to God.

2. The argument from the nature of efficient cause. We know of no case, indeed there couldn't possibly be a case, in which something is the cause of itself, for nothing can be prior to itself. But, again, this chain of 'cause' and 'what has been caused' cannot go on to infinity. 'It is necessary to admit a first efficient cause, to which everyone gives the name of God.'

With Hawking's no-boundary proposal it seems we have a case in which something could be the cause of itself without existing prior to itself.

3. The argument from possibility and necessity. In nature it appears that all things have the possibility either of being or not being. Anything that has the possibility of not being must at some time not be – so it is impossible for things always to have existed. It follows that there was a time when nothing existed, everything was only a possibility. Things cannot begin to exist on their own. Moving from possibility to existence requires something to cause that change, and obviously that change has taken place. There-fore, it must *not* be true that *all* things at one time did not exist and were merely possible. There must be something to cause that change, and this something's existence must not be merely possible but necessary in and of itself, without being caused by something else. 'This all men speak of as God.'

We have seen a proposal for dealing with just this dilemma in the suggestion that nothingness is unstable and tends to decay into something. What is necessary in and of itself, in that proposal, is that nothingness be unstable to that degree. But Polkinghorne has pointed out that we don't get that instability 'for nothing'.[22] Hawking echoes Aquinas when he asks: 'What is it that breathes fire into the equations and makes a universe for them to describe? . . . Is the unified theory so compelling that it brings about its own existence?'[23] To use words closer to Aquinas's, what is it that changes the possibility of anything at all existing, into something that really exists? It appears to be a step of insurmountable proportions, making the details of existence, once it occurs, seem insignificant by comparison.

4. The argument from the gradation to be found in things. There is gradation in all things – more good or less good, more noble or less noble, more hot or less hot – which implies that there is something which is the maximum against which this 'more' or 'less' is judged. Without the maximum standard in such things as 'being, goodness, and every other perfection' the gradations could not exist; they are in effect caused by the existence of the maximum. 'Therefore there must also be something which is to all beings the cause of their being, goodness, and every other perfection; and this we call God.'

This argument was echoed in Chapter 3 in our discussion of the moral arrow – what is it that defines the directions 'good' – 'evil' or 'truth' – 'falsehood' in the universe? – and in our discussion of God as the ultimate self-confirming hypothesis in Chapter 6.

5. The argument from the governance of the world. Things which lack knowledge, natural bodies for example, nevertheless evidently act for an end, acting always, or nearly always, in a way which will lead to the best result. This can't be mere luck; nor, lacking any knowledge, could these things be doing this on their own volition. They must be directed by something which does have knowledge and intelligence. 'Therefore some intelligent being exists by whom all natural things are directed to their end; and this being we call God.'

In spite of this prodigious attempt to find rational arguments for the existence of God, Aquinas himself insisted that arguments for faith by means of human reason cannot succeed in proving what belongs to faith. He thought human reason indispensable, on the other hand, for arguing from articles of faith to other truths.

Was this a case of having to believe in advance in order for the other evidence and arguments to have meaning? It was, and Aquinas knew it was. However, he saw in this a direct similarity between his 'science' and other sciences. In fact, what Aquinas meant by a 'science' was a body of knowledge that has 'first principles'.

'As the other sciences do not argue in proof of their principles, but argue from their principles to demonstrate other truths in these sciences, so this doctrine does not argue in proof of its principles, which are the articles of faith, but from them it goes on to prove something else.'[24] This is an insight we seldom bring to a controversy which pits the value of scientific knowledge and evidence against the value of religious knowledge and evidence. We tend to allow science to rely on its underlying assumptions and insist only that all else follow logically from there. Very seldom do we ask science to defend those assumptions. We do not, on the other hand, tend to allow religion to rely even on its most basic principle – that there is a God – and let all else follow logically from there. Religion is repeatedly called back to defending that basic principle. Is it any wonder that science and religion usually seem to be arguing past one another?

One of the themes of this book has been that the assumptions which underlie science are as unprovable and logically indefensible as those which underlie religion – in fact they are largely the same assumptions. One assumption which we touched on only briefly in Chapter 2, but which has emerged in later chapters as an underlying assumption of both science and religion, is the authenticity of human experience. In this chapter we have stressed the importance of human experience in deciding whether there is a God, and science likewise has relied heavily on human experience as opposed to pure reason since the emergence of the scientific method. Yet the authenticity of human experience is also only an assumption. We have no proof of that authenticity. Nevertheless we *do* assume that our experience is valid evidence. And that leaves us with such mysteries as the news that some time after Aquinas had laid all these arguments out in his *Summa Theologica* he had an experience of the presence of God which led him to speak of all this previous intellectual effort as resembling mere 'straw'.[25]

THE ARGUMENT FROM EXPLANATORY POWER

We learned in our previous discussions of science that one strong argument for the validity of a theory is its ability to make sense of data which has previously seemed confusing and unexplainable. Can we find such arguments in defence of religion?

One is the 'argument from morality', that we have in religion a strong description and rationale for human nature, the human condition, and the orientation of good, evil, innocence, and guilt in which we find ourselves. The argument from morality also has it that in religion we have an unparalleled moral code, as well as a viable way of getting out of the moral dilemma we find ourselves in when we can't live up to that code. Though there were other sophisticated ancient moral codes; and though there is other literature besides the Bible – Shakespeare and the Greek dramatists, for instance – which depicts human nature with similar sensitivity, humour, and accuracy; and though the science of the mind, psychology, offers an additional way to deal with our guilts and other mental difficulties; none of these offers such a deep and comprehensive picture as the Bible does.

Aquinas, as we saw earlier, pointed to a 'gradation in all things' and said that without a maximum standard in such things as 'being, goodness, and every other perfection' these gradations couldn't exist. C. S. Lewis echoed Aquinas in insisting that this maximum standard can be no other than God, that it is very difficult to explain how the good–evil compass exists at all in any form in the universe (and we don't seem able to escape the impression that it *does*) without positing a God as the ultimate standard.

Immanuel Kant, the eighteenth-century German philosopher best known for his *Critique of Pure Reason*, is of all philosophers probably the least possible to discuss adequately in anything short of book length. Nevertheless, let us attempt a summary. Kant argued for faith on the grounds of moral value. Kant, like Hume, pointed out the limits of reason in the search for truth, and he demonstrated that rational proofs could not be found to show that there was a God who was the First Cause of the universe. Kant lacked experiential evidence of God. Furthermore, he attacked the notion that the rationality and artistry of the universe are pointers to the existence of God, on the grounds that they are useless to demonstrate or explain the moral wisdom that we attribute to God.

According to Kant it is this moral wisdom which makes faith in God reasonable and not mere uncritical acceptance of unproved propositions. His argument goes like this. We have a moral duty to promote the highest good. Saying that we 'ought' to achieve this highest good implies that we 'can' achieve it. Nevertheless we know that we never do achieve it. Virtue we might be able to achieve, but virtue does not ensure happiness, and 'highest good' means happiness as well as virtue. The requirement of 'highest good' can be met only by postulating a moral and rational being who created the universe and sustains it, and who has the power to see that happiness is proportional to virtue.

It might appear that, although Kant said that truth can't be established by reason, that is precisely what he was trying to do. But he didn't claim that he had by this line of thought established objective truth. He argued only that he could see no other way to account rationally for our moral experience than to postulate that there was a rational and moral creator and sustainer of the universe, and also that there was freedom of will and immortality of the soul. The moral wisdom of God, additionally, seemed to Kant the only rational basis for personal conduct.

For Kant, the moral wisdom of God could not be proved and had to remain largely outside our understanding. The presence of evil and suffering couldn't be explained away, and all attempts to do so were weak and inadequate, even offensive. Better Job's attitude of submission without understanding.

Although Kant insisted that there is no way of proving whether or not God exists by means of speculative reason or science or by the argument of moral wisdom, he also rejected the notion that we need to know whether or not God exists before deciding whether or not to take a religious stand. Rather, he would have us know that we do not know, and concede that the idea of a God, while not provable, is defensible because of the results of accepting the moral wisdom of God.

We should point out that though many theologians have criticized Kant for relegating Christ to the status of a moral teacher, that is a rather simplistic interpretation of Kant. What emerges more strongly from his writing is not the idea that we must abandon the notion that there is a God or decide that the moral precepts are the only part of religion we can preserve, but rather that we must abandon attempts to prove or disprove God, and, even more, to manipulate God, as though he were an object. Accepting the moral attitude – along with Christianity's encouragement that, when we are unable to live up to the pure and uncompromising demands of that morality, 'if we act as well as lies in our power, what is not in our power will come to our aid from another source, *whether we know in what way or not*'[26] (author's emphasis) – amounts to a faith beyond the espousal of a moral code. That this was the most reasonable way to proceed seemed to Kant a sufficient argument for faith.

Today Kant would have to face those who would argue that our moral arrow and the reason why we might seem to live more successfully within a certain moral order come from evolutionary and societal history and have no relevance to the question whether there is a God. Behaviour that gave a survival advantage and later was reinforced when it was imposed by societies seeking to preserve themselves has been wired into our psychological make-up as 'moral behaviour'. Other possible configurations of good vs. evil have not. Kant would also have to contend with those who see the moral code of the Bible as outdated, and those who view biblical values as superseded by 'human' values, values we are capable of choosing for ourselves without any help from God and without any other explanation or rationalization. Modern philosophical descendants of Kant might counter-argue that if we take a step back we find that all evidence pointing to an evolutionary explanation or a 'human-values' explanation can serve equally well, perhaps better, as evidence that the good–evil arrow in the universe – and human longing to achieve true good and inability to do so – are as fundamental and binding as any physical law. The difficulty of explaining this situation, except by postulating the existence of God and accepting the biblical description of the God–human relationship, may be even greater than Kant thought.

Another argument from explanatory power is offered to counter a viewpoint we saw earlier which insisted that, judging by results of belief for individuals and societies, religion has little to recommend it. This argument also counters the claim that the existence of evil and outright absurdity is evidence that there is no God, or that, if there is, God must be, in the words of playwright Tennessee Williams, 'a senile delinquent'.[27] The argument is that the Bible doesn't promise us, either as individuals or as society, the rosy circumstances, the absence of evil, or the rationality that sceptics are looking for. The New Testament in particular, rather than predicting that sweetness and light for humankind will result from the fact that there is a God or from the good works of those who believe in God, insists repeatedly to the contrary: in spite of good efforts and short-term gain, until Christ returns, the world will increasingly be beset by relentless evil, hatred, suffering, injustice, and violence, as well as confusion wrought by those who falsely or mistakenly claim to be acting in the name of God or even insist that they themselves are Christ. It has so far been a disturbingly accurate prediction. We see this apocalyptic vision played out all too regularly on the television news.

When it comes to individuals, the Bible doesn't give us any reason to expect that those who believe in God will all be virtuous. Quite the contrary. The promise is only that they will be forgiven. Nor, despite the snake oil dispensed by some television evangelists, does either testament lead us to suppose that believers will all be happy, healthy, and successful by this world's standards.

Judaism and Christianity have from ancient times wrestled with the problem of evil on every scale. They have insisted that evil is something which human endeavour alone will never conquer or human reason completely understand, and at the same time assured us that we are absolutely right to despise it and try to combat it as best we can.

Hence it is inappropriate to discount the validity of biblical evidence or religion in general on the grounds that all is not well with the world, that evil is rampant, and that God and belief in God have not transformed this planet into a Garden of Eden or believers into earthly angels. The biblical claim is only that God will some day cause a transformation something like that to occur. 'Something completely different' at last! We have yet to be able to evaluate that evidence, and it does seem that we have been waiting a very long time. Meanwhile, those who believe in the biblical God point out that the biblical description of the world situation is, sadly, on target, and that the biblical explanation for our human predicament and the problem of evil – though we may not like or accept that explanation – fits.

THE ARGUMENT FROM NATURE

We asked at the end of Chapter 3 whether anywhere in nature as we have come to know it, 'In reason's ear' we can hear the words clearly spoken 'The hand that made us is divine.' Nowhere have we found that statement made by nature in a way that proves unequivocally the existence of God. But the arguments from design and rationality have not disappeared in these last decades of the twentieth century. For whatever cause, valid or not, we still find inescapable the impression of a Mind behind the laws of this universe or inherent in them. Nowhere have we discovered mindlessness, not even in the unpredictable systems studied by chaos scientists, where there is mysterious, beautiful pattern and structure. Even our depressing vision of a universe running down in an inexorable increase in entropy is being replaced by a picture that also shows us self-organization on every scale.

Human beings have come full circle more than once, trying to find evidence of God in nature. Some of our most distant ancestors found God or gods in the chaos of nature – the fearfully unpredictable and arbitrary – gods to be appeased, not loved or understood. Later the rationality, the pattern, the legality of the universe seemed eloquent testimony of the Mind of God. Then we found that the pattern was so strong, the systemization so universal, that the universe hardly needed a God. Everything operated like clockwork, a clockwork perhaps even capable of having invented itself. It seemed that if there were a God we must return to the more ancient way of looking for him in the places where the clockwork broke down or skipped a beat, in the unpredictable, the places science chose to ignore – the gaps. However, science turned its searchlight on those areas too, and it looked as though there soon would be no gaps left in which we could conceive of God existing and exerting his power in the universe.

Now we discover to our surprise that the clockwork breaks down nearly everywhere. Predictable systems are the exception, not the rule. In fact, we hardly find them at all except for rather specialized situations, and then they are predictable only to a limited extent. Yet we haven't discovered irrationality, and the fact that we haven't – that the universe *is* rational – suddenly begins to look as though it may be as difficult to explain as our ancestors thought it was. What right have we to expect that the universe should have organized itself into galaxies, stars, and planets; that life on this earth would have organized itself into ecosystems, and animals and human beings into societies? There are probabilities, but by some calculations those probabilities are vanishingly low. It begins to look as though the universe couldn't exist without a mysterious tendency to organize, though never to the extent that the mechanists and the determinists hoped.

It was the tendency toward simplicity and linearity which in the past made nature accessible to our minds. We are only now beginning to understand that a deeper understanding of nature – dare we even call it a deeper accessibility – lies in recognizing that much of it will never reduce to simplicity and linearity, and in attempting to find out why any of it should.

Joseph Ford gives an eloquent illustration of this dynamic broader symmetry between chaos and order in his 1989 article 'What is Chaos That We Should Be Mindful of It?', where he speaks of chaos as 'dynamics freed from the shackles of order and predictability', permitting systems 'to explore at random their every dynamical possibility' – a 'cornucopia of opportunities'. Ford points to evolution as an example of how nature uses the randomness of chaos to generate order:

> In setting up the scheme humans call evolution, nature wished to insure in perpetuity the survival of life forms against every possible variety of natural catastrophe; in addition, nature wished to encourage the expansion of life into every possible ecological niche no matter how harsh or specialised. In principle, nature could have written a deterministic program to cope with the temporal unfolding of exceedingly complex, almost random patterns of life affecting events.

Also in principle, God could have written such a program, or God could have set it all up piece by intricate piece, putting everything in by hand – so that each infinite detail of creation would be an arbitrary element. Ford goes on:

> Instead, nature chose a highly effective technique which uses randomness to defend against the unexpected. Specifically, nature uses random mutations to provide the wide variety of life forms needed to meet the demands of natural selection. In essence, evolution is chaos with feedback. Random mutations alone would correspond to nature indifferently rolling unbiased dice, but the added feedback of natural selection and survival of the fittest, in effect, biases the dice so that, over many rolls, life forms not only survive, they improve, probability one.[28]

Ford explains that when he says an event occurs 'probability one', he means its occurrence is overwhelmingly likely.

Who or what is 'nature'? Did this system itself evolve from less effective systems? Can it all be explained as the outcome of probabilities – a statistical context mysteriously laid down at the origin of the universe or even before which rendered sentient life inevitable? Richard Dawkins implies that it can. Other scientists say DNA 'should never have arisen'. *Is* there a mysterious organizing principle at work, one which science will discover – or perhaps one which is beyond the power of any scientific Theory of Everything to explain?

276

Which then speaks more eloquently on behalf of there being a God – the pattern and rationality of the universe, or what seems unexplainable and arbitrary? Perhaps it is a greater symmetry, the way both chaos and order are enmeshed and intertwined, that is better evidence of an infinitely superior mind, willing to allow and encompass risk and freedom in a way that humans, with our careful parsimony, find fearful. We do know that on the level of human creativity – in art and music and dance and literature – genius takes enormous risks.

In this context, even for some who do not believe in God, the attempts we have studied in previous chapters which try to explain the universe without God begin to look slightly contrived and self-serving, and the older, simpler, more mysterious explanation, 'There is a God', less so.

But we have not discovered proof of God in nature to show our hypothetical alien. All efforts to discover such proof have failed. Newton thought he had found clear evidence of providence in the replenishing of solar and planetary matter by means of comets, but that idea was defeated at the end of the eighteenth century when La Place and La Grange showed that irregularities induced in planetary orbits could be self-correcting. Paley thought he had found clear evidence of providence in the complex designs of nature, but Darwin discovered evolution. It seems we come near catching the creator red-handed, but he slips through our fingers – as happens in some mystery stories where the detectives, certain they are about to apprehend the culprit at last, instead find only a clever mechanical device left to deceive them. If, from the clever device, we cannot even prove there is a culprit, does that indicate that there is none, or that the culprit is surpassingly clever?

THE ARGUMENT FROM AVAILABILITY

In his lecture 'The Invincible Ignorance of Science', Brian Pippard relegates himself and many of his friends to the status of craftsmen in the City of God for ever. T. S. Eliot in *The Cocktail Party* presents almost that same view of the universe. There are the saints, and then there are the rest of us. Yet as we have seen, if there *is* knowledge to be had about God, and if the Judaeo-

Christian religion is at all right about that knowledge, it is the most universally available of all knowledge. It is distressingly egalitarian. The elderly lady with blue hair and pink curlers and appalling taste can 'know the Lord' just as well as St Augustine or the Archbishop of Canterbury, and she may be infinitely nearer the Mind of God than Stephen Hawking is. That is a dogma we might like to despise, even though we might not want to admit that we are so elitist. But there is hardly any religious claim stronger than that the answer to the question 'Is there a God?' is universally available to those whose desire to know goes beyond mere intellectual curiosity. That is, of course, the argument which we have chosen not to offer our alien friend. In so choosing, we may have deprived the alien of the only really valid evidence for God.

We don't know how large a proportion of the significant evidence about the universe is excluded by science. Perhaps hardly any. Perhaps so great a proportion that any body of knowledge which excludes it is hardly more than a caricature. Perhaps something in between – so that science finds truth but not the whole truth. Polkinghorne has compared the exclusion of all private evidence to studying the universe using optical telescopes without being able to use a radio telescope. Those who wish that science could accept what is now inadmissible evidence are unable to suggest how that might be done.

More perhaps than anything else, what sets religious evidence apart is not really its inability to be corroborated or its failure to be 'public evidence' by the standard of science, but its very richness. Religion and science, as well as art, music, and literature – all are rooted in human experience. There is of course a commonality to that experience, but at the most fundamental level human experience can never be shared, and all knowledge enters first on that level. Science is eager to process this knowledge into public knowledge by moving on as rapidly as possible to comparison, argument, and consensus – and that processing has served us magnificently for learning about the physical universe. However, some scepticism might be the better part of wisdom when it comes to the use of this same process as the ultimate arbiter of the validity of *all* human experience.

8

THEORY OF EVERYTHING . . .
MIND OF GOD

Then we shall all, philosophers, scientists, and just ordinary people, be able to take part in the discussion of the question of why it is that we and the universe exist. If we find the answer to that, it would be the ultimate triumph of human reason – for then we would know the mind of God.

STEPHEN HAWKING[1]

If you accept my words and store up my commands within you, turning your ear to wisdom and applying your heart to understanding, and if you call out for insight and cry aloud for understanding, and if you look for it as for silver and search for it as for hidden treasure, then you will understand the fear of the Lord and find the knowledge of God. For the Lord gives wisdom, and from his mouth come knowledge and understanding.

PROVERBS[2]

If there are ultimate answers, whatever conclusions we reach in this book won't change them. No depth of human need, no radiant faith, no convincing argument or intellectual exercise can create a real God, if there is no God. No honest agnosticism, no stark atheism, no brilliantly successful scientific explanations, no inconsistency between science and belief can cause God not to exist, if there is a God. We do not decide these things. Can we ever *know* them? Human intellectual endeavour has not yet built the ladder that can take us to the ultimate answers. Arguably it never will.

We are back where we began. I see across the room from me my grandparents' wooden chair. Is it here, with myself, that my quest for knowledge both begins and ends? However far afield the journey takes me and however public it becomes, does it only lead back to my mind's-eye view of the universe? After all, who or what besides myself will decide what *I* accept as truth? I know that if there are ultimate answers I'm surely not the court of last appeal as to what those answers are, but here in my study, on this human level, for myself, it appears I am. Does it matter very much what I decide? Not to science. But religion would have me think that the decisions of this private court when it comes to whether or not I will believe in God are of inestimably great significance.

In the intellectual exploration we've undertaken in this book, we've found that the choices of what we should accept from science and from religion are not as stark as they have often been depicted. Our journey has several times taken us past potential flash points – the discussion of evolution and a purposeful God, for example – where we had been led to expect grave and irreconcilable conflict, but where such conflict didn't arise. However, all is not in agreement.

With some of the science, we've been talking about scientific findings and well-established knowledge, in other instances about unproved theoretical assertions and speculations. Some of the religious beliefs we've discussed are central to religious faith, others are only peripheral to the question 'Is there a God?' and even to the question 'Is there a personal God who interacts with his creation?' Among all this we are certainly able to choose from 'science' and 'religion', as from Column A and Column B in a restaurant, so as to end up with a serious though perhaps not terminal case of logical indigestion.

Those who believe that God literally plucked Eve from Adam's side have a conflict with those who assert that human beings of both sexes evolved over a lengthy period of time from less complex forms of life. It would be difficult to hold both beliefs simultaneously except possibly by recourse to the complementarity we discussed in Chapter 6. But can this really be called a conflict between science and religion, when the majority of devoutly religious persons do not feel it necessary to believe that God literally plucked Eve from Adam's side? Arguably we have done no more than pick two items from Column B. Those who believe that God stopped the sun for Joshua have a conflict with those who assert that we know the laws which govern change in this universe well enough to conclude unequivocally that this event couldn't happen, and who also assert that God never breaks scientific laws. It would be difficult to hold both beliefs simultaneously unless, again, we were to insist that it is sometimes more realistic to live with a contradiction than to try to resolve it prematurely. But is this really a conflict between religion and science, when most religious people think the Joshua story is possibly a legend which grew up around an early hero, and find that thought in no way deleterious to belief in God? The belief that Christ rose from the dead, a belief far more central to Christian belief than the Joshua story is to either Judaism or Christianity, conflicts with the insistence that such a reversal of the normal biological processes is not possible. But is that really a conflict between science and religion, when some scientists think God sometimes sets aside his laws, and when one scientist I consulted in connection with a scientific theory in this book told me, 'No, I don't think God breaks his own laws, but I also don't think I *know* the fundamental laws which God doesn't break, simply by virtue of knowing some approximations based on what normally happens'? We perhaps have merely chosen two items from Column A.

We can hardly deny that there is a serious conflict between belief that science will eventually show that there is no God and belief that there is a God. But is this a conflict between 'science' and 'religion'? Surely we can give that dispute no more dignity than to call it either excessively optimistic atheism, wearing a cloak of science, vs. religion; or, worse, two blind faiths confronting one another not only on imaginary horses but on an imaginary battlefield. It is possible to go on contriving conflicts, compromises, and happy resolutions to our hearts' content, and to arrange and rearrange our knights on the field so as to bring about outright carnage, decorous games, or a beating of swords into ploughshares. It behoves all of us, when someone announces

'Science and religion are in conflict', to ask what specific items this person has chosen from Column A and Column B. 'Well . . . you know . . . Galileo' will not do.

Where have we arrived at the end of seven chapters? Joseph Ford has said: 'More than most, [scientists] are content to live with unanswered questions.'[3] One of the questions science hasn't answered and may never be able to answer – let none of us assume otherwise – is whether there is a God. We have *not* been able to say that it requires double-think or other intellectual dishonesty to have great faith in science as we know it at the end of the twentieth century and also to believe in God – even a personal and intervening God.

But why should anyone think such a combination of faiths might be necessary, or indispensable on a quest for fundamental truth? There are two reasons for thinking it might be. One would be to have first-hand, experiential evidence of God which was personally convincing. The second is because to dismiss belief in God summarily is to pass premature and unwarranted judgement on the sanity, honesty, and intelligence of a vast number of our fellow human beings who claim to have such experiential evidence, many of them the same persons we do trust implicitly when it comes to other matters. It ill becomes any of us to take the attitude that all evidence for God is false evidence, beneath consideration, simply by virtue of its being evidence for God, or even by virtue of its being outside the purview of science. Such attitudes are taken, sometimes in the name of science, but in truth this sort of attitude *is* intellectual dishonesty. Our most reputable scientists, whatever sins of arrogance they may occasionally commit, do not really declare that what they don't know isn't knowledge or that what they haven't experienced isn't experience.

Science leads us to hope that complete understanding is potentially within the grasp of human collective reason, but science is not overly confident of finding it. 'I think I may find out "how" but I'm not so optimistic about finding out "why." If I knew that, I would know everything important', says Stephen Hawking.[4] John Barrow writes: 'There is no formula that can deliver all truth, all harmony, all simplicity. No Theory of Everything can ever provide total insight. For, to see through everything would leave us seeing nothing at all.'[5] But St Paul wrote: 'Now we see but a poor reflection as in a mirror; then we shall see face to face. Now I know in part; then I shall know fully, even as I am fully known.'[6] Religion is far more optimistic than science that in some manner beyond our present concept of human reason, we can know 'everything important'. Perhaps the most significant difference between science and religion is that science thinks that on this quest we are entirely on our own. Religion tells us that although we who seek the truth may ride imaginary horses, Truth also seeks *us*.

—

NOTES

CHAPTER 1

1. Quoted in James R. Moore, 'Charles Darwin Lies in Westminster Abbey', *Biological Journal of the Linnean Society* (1982): 102.
2–5. Ibid., 103.
6. Proverbs 3:13.
7. Proberbs 2:5, 6.

CHAPTER 2

1. Published by Macmillan Company and Cambridge University Press, 1928.
2. Stephen W. Hawking, *A Brief History of Time: From the Big Bang to Black Holes* (London: Bantam Press, 1988): 174.
3. Brian Pippard, 'Eddington's Two Tables', *Great Ideas Today: 1990* (London: Encyclopedia Britannica, 1990): 316.
4. Ibid., 312.
5. Quoted in 'Edison Enlightens', *Uncle John's Third Bathroom Reader* (New York: St Martin's Press, 1990): 161.
6. *The Four Quartets*: 'East Coker'.
7. 'Physics and Reality', *Journal of the Franklin Institute*, 221 (1936): 349.
8. *QED: The Strange Theory of Light and Matter* (Princeton: Princeton University Press, 1985): 4.
9. Quoted in Bryan Appleyard, 'Master of the Universe: Will Stephen Hawking Live to Find the Secret?', *Sunday Times*, 3 July 1988.

CHAPTER 3

1. 'The Value of Science' in *The Foundations of Science* (New York: Science Press, 1907): 318.
2. Quoted in John Winokur, *Einstein: A Portrait* (California: Pomegranate Art Books, 1983).
3. Quoted in Michael Harwood, 'The Universe and Dr Hawking', *New York Times Magazine*, 23 January 1983: 57.
4. 'The Value of Science': 199.
5. *Science and Religion: Some Historical Perspectives* (Cambridge: Cambridge University Press, Cambridge History of Science Series, 1991): 257.
6. *The Physicist's Conception of Nature* (London: Greenwood Press, 1958): 24.
7. Quoted in Kitty Ferguson, *Stephen Hawking: Quest for a Theory of Everything* (London: Bantam Press, 1992): 30.
8. *Theories of Everything: The Quest for Ultimate Explanation* (Oxford: Clarendon Press, 1991): 125.
9. London: Freeman Cooper, 1969.
10–11. *Dreams of a Final Theory: The Search for the Fundamental Laws of Nature* (New York: Pantheon Books, 1992): 123.
12. Ibid., 125.
13. Ibid., 126, 127.
14. Quoted by George Bruce Halsted in one of the introductions (titled 'Henri Poincaré') to Poincaré, *The Foundations of Science* (New York: Science Press, 1929): x. Halsted gives the source of the quote as Poincaré, *Électricité et Optique*, 1901.
15. *The Value of Science*: 353.
16. Jagdish Mehra told this anecdote at a dinner given for Dirac on his seventieth birthday, in Trieste, Italy, in September 1972. Printed in Jagdish Mehra (ed.), *The Physicist's Conception of Nature* (Norwell, Mass.: Kluwer Academic Publishers, 1973).
17. 'The Evolution of the Physicist's Picture of Nature', *Scientific American*, May 1963: 47.
18. *A Mathematician's Apology* (Cambridge: Cambridge University Press, 1940): 25.
19. Weinberg, *Dreams of a Final Theory:* 98.
20. Murray Gell-Mann, lecture.
21. Albert Einstein, letter to Ernst Strauss, reprinted in B. G. Kuznetsov (trans. H. Fuchs), *Einstein: Leben, Tod, Unsterblichkeit* (Basel: Birkhauser, 1977): 285.
22. *A Mathematician's Apology*: 70.
23–24. Quoted in Jonathan Powers, 'Did God have any Choice in

the Creation of the world?', *Symposium: Hawking's 'History of Time', Re-considered, The Cambridge Review*, March 1992: 13.

25. 'Consistency and Completeness: A Résumé', *American Mathematical Monthly* 63 (1956): 295–305.
26. *Pi in the Sky: Counting, Thinking, and Being* (Oxford: Clarendon Press, 1992).
27. *Theories of Everything:* 38.
28. Quoted in John Tierney, 'Subrahmanyan Chandrasekhar: Quest for Order', in Allen Hammond (ed.), *A Passion to Know: Twenty Profiles in Science* (American Association for the Advancement of Science, 1984; New York: Scribner, 1985): 6.
29. The story about Chandrasekhar and Eddington is told ibid., 2–5.
30. Quoted ibid., 3.
31. Quoted ibid., 4.
32. 'A Killer Returns', *Newsweek*, 30 November 1992: 39.
33. *Superforce: The Search for a Grand Unified Theory of Nature* (London: Heinemann, 1984): 47.
34. Translation from the Dutch, John Bowden, *More Things in Heaven and Earth: God and the Scientists* (London: SCM Press, 1991): 42.
35. Deuteronomy 6:16, as quoted by Christ in Matthew 4:7.
36. *Dreams of a Final Theory*: 247.
37. London: W. W. Norton & Co., 1978; reprinted 1992: 9.
38. Einstein in a letter to Willem de Sitter, quoted ibid., 21.
39. *Time*, 30 December 1974: 48.
40. *A Brief History of Time*: 141.
41. *The Blind Watchmaker: Why the evidence of evolution reveals a universe without design* (London: W. W. Norton & Co., 1986): 255.
42. *Dreams of a Final Theory*: 188.
43. 'Muddling to Discovery', *Newsweek*, 24 August 1992: 52.
44. *A Brief History of Time*: 174.
45. Invented by mystery writer John Dickson Carr.
46. ABC *20/20* Broadcast, 1989.
47. Quoted in Paul C. W. Davies, *The Mind of God: Science and the Search for Ultimate Meaning* (London: Simon & Schuster, 1992): 223.
48. *Surprised by Joy: The Shape of my Early Life* (London: Harcourt, Brace and World, 1956): 191.
49. Closing line of the hymn 'The spacious firmament on high', poem by Joseph Addison (1672–1719), loosely paraphrasing Psalm 19:1–6.

CHAPTER 4

1. Dr Seuss, *Horton Hears a Who* (New York: Random House, 1954).
2. Quoted in Jastrow, *God and the Astronomers*: 18. The story was told to Jastrow by John Hall, at one time Director of Lowell Observatory at Flagstaff, who heard it from John Miller.
3. Letter to Willem de Sitter, quoted ibid., 21.
4. Translated by Betty H. Korff and Serge A. Korff, *The Primeval Atom* (copyright 1950), reprinted in Timothy Ferris, *The World Treasury of Physics, Astronomy, and Mathematics* (New York: Little, Brown & Company, 1991: 360.
5. The discussion here uses ten billion years for the sake of illustration. The age of the universe is presently estimated at between ten and twenty billion years. Light from very distant quasars was emitted when the universe was approximately 6 per cent of its present age.
6. *God and the Astronomers*: 107.
7. 'The Origin of the Universe', lecture delivered at the Three Hundred Years of Gravity Conference in Cambridge, June 1987. Reprinted in Stephen W. Hawking, *Black Holes and Baby Universes, and other Essays* (London: Bantam Press, 1993): 91.
8. 'The Edge of Spacetime', in Paul C. W. Davies (ed.), *The New Physics* (Cambridge: Cambridge University Press, 1989): 67.
9. New York: Viking, 1986: 93.
10. Ibid., 92.
11. Papal address to the Conference of Astronomical Cosmology, at the Vatican, 1981.
12. Quoted in Roy E. Peacock, *A Brief History of Eternity: A Considered Response to Stephen Hawking's 'A Brief History of Time'* (London: Monarch Publications, 1989): 93.
13. 'The Edge of Spacetime': 68.
14. 'Einstein's Dream', lecture delivered at the Paradigm Session of the NTT Data Communications Systems Corporation in Tokyo, July 1991. Reprinted in *Black Holes and Baby Universes*: 83.
15. Quoted in Ferguson, *Stephen Hawking*: 119.
16. Quoted in Jerry Adler, Gerald Lubenow, and Maggie Malone, 'Reading God's Mind', *Newsweek*, 13 June 1988: 59.
17. *Listen, There's a Hell of a Good Universe*.
18. See Peter Coveney and Roger Highfield, *The Arrow of Time* (London: W. H. Allen, 1990): 181.
19. Quoted in H. R. Pagels, *Perfect Symmetry* (New York: Simon & Schuster, 1985): 316.
20. *Summa Theologica*, Part 1, Question 2, Third Article (translator

not cited), in Steven M. Cahn (ed.), *Classics of Western Philosophy* (Cambridge: Hackett Publishing Co., 1977, 3rd edition, 1990): 401.

21. *A Brief History of Time*: 174.
22. Genesis 1:1.
23. *A Brief History of Time*: 9.
24. *The Emperor's New Mind: Concerning Computers, Minds, and the Laws of Physics* (Oxford: Oxford University Press, 1989): 94, 95.
25. *A Brief History of Time*: 174.
26. *Theories of Everything*: 185.
27. Halley Lecture, Oxford, June 1989.
28. Quoted in Appleyard, 'Master of the Universe'.
29. *The Mind of God*: 151.
30. 'The Mind of God?', *Symposium: Hawking's 'History of Time' Reconsidered, The Cambridge Review*, March 1992: 1.
31. Personal letter to the author.
32. Neither Hartle nor Hawking has used the words 'First Cause' in reference to the no-boundary proposal.
33. 'Master of the Universe: Stephen Hawking', BBC broadcast, 1989.
34. Don N. Page, 'Hawking's Timely Story', *Nature*, 332, 21 April 1988: 743.
35. Colossians 1:16–17.
36. *The Mind of God*: 68.
37. Augustine of Hippo, *Confessions*, Book XI, 'In the beginning God created . . .' trans. F. J. Sheed, in Cahn (ed.), *Classics of Western Philosophy*: 360.
38. See Page, 'Hawking's Timely Story': 743.
39. 'My Position', a talk at Gonville and Caius College, Cambridge, in May 1992; reprinted in *Black Holes and Baby Universes*: 46.
40. *A Journey into Gravity and Spacetime* (New York: W. H. Freeman & Co., 1990): 3.

CHAPTER 5

1. Quoted in A. Moszokowski, *Conversations with Einstein* (New York: Horizon, 1970).
2. *The God Particle: If the Universe is the Answer, What is the Question?* (New York: Houghton Mifflin Company, 1993): 1.
3. 'Master of the Universe', BBC broadcast, 1989.
4. From an essay by Albert Einstein written in 1939 and reprinted in his *Out of My Later Years* and in Ferris, *World Treasury of Physics, Astronomy, and Mathematics*: 835.

5. *A Brief History of Time*: 174.
6. The tea-kettle story comes from Polkinghorne's book *One World: The Interaction of Science and Theology* (London: SPCK, 1986): 62.
7. *The Blind Watchmaker:* 4.
8. William Paley, *Natural Theology – or Evidences of the Existence and Attributes of the Deity Collected from the Appearances of Nature*, 1802. Reprinted by St Thomas Press, Houston, Texas, 1972. Quoted in Dawkins, *The Blind Watchmaker*: 4.
9. p. 213.
10. *The Blind Watchmaker*: 4.
11–12. Ibid., 288.
13. For Dawkins' program, see ibid., Chapter 4, 43, etc.
14–15. Ibid., 141.
16. p. 409.
17. *A Brief History of Time*: 133.
18. p. 53.
19. Quoted in David H. Freedman, 'Maker of Worlds', *Discover*, July 1990: 49.
20. 'Master of the Universe', BBC broadcast, 1989.
21. Psalms 8: 3, 4.
22. Psalms 46: 1–3.
23. See Brooke, *Science and Religion* : 52.
24. Quoted in Peacock, *A Brief History of Eternity*: 22 and 38; also Colin Humphreys, 'Can Science and Christianity Both Be True?' in R. J. Berry (ed.), *Real Science, Real Faith* (Eastbourne: Monarch, 1991): 116.
25. Quoted in Owen Gingerich, 'Let There Be Light: Modern Cosmogony and Biblical Creation', Roland Mushat Frye (ed.), *Is God a Creationist?* (New York: Charles Scribner's Sons, 1983); reprinted in Ferris, *The World Treasury*: 393.
26. p. 447.
27. Personal letter to author.

CHAPTER 6

1. Stephen W. Hawking, 'In Defence of "A Brief History"', *Cambridge Review* (March 1992): 16.
2. Ecclesiastes 1:9.
3. Often repeated promise on *Monty Python's Flying Circus*.
4. Sir Nevill Mott, 'Christianity Without Miracles?', Nevill Mott (ed.), *Can Scientists Believe?: Some Examples of the Attitude of Scientists to Religion* (London: James & James, 1991): 4, 5.

5. Joshua 10:12–14.

6. Sir Arthur Conan Doyle, 'Silver Blaze', *Memoirs of Sherlock Holmes*.

7. *Science and Providence: God's Interaction with the World* (London: SPCK, 1989): 27, 28.

8. *The Blind Watchmaker*: 159, 160.

9. 'What is Chaos That We Should Be Mindful of It?', in Paul C. W. Davies (ed.), *The New Physics* (Cambridge: Cambridge University Press, 1989): 348.

10. 'The Promise of Chaos: An Interview with Georgia Tech Physics Professor Joseph Ford', in *Georgia Tech Research Horizons*, Spring 1988: 14.

11. 'What is Chaos?': 351.

12. James Clerk Maxwell, Essay: 'Does the progress of Physical Science tend to give any advantage to the opinion of Necessity (or Determinism) over that of the Contingency of Events and the Freedom of the Will?', written in 1873; reprinted in Lewis Campbell and William Garnett, *The Life of James Clerk Maxwell* (London: 1882; New York: Johnson Reprint Corporation, 1969): 440, 442.

13. p. 206.

14. London: Heinemann, 1985: 271.

15. 'What is Chaos?': 352.

16. Bereshit Rabbah, Ch. 5 (collection of dicta of Talmudic sages, edited in the fifth century; English translation Sancino, 1939). Quoted in Cyril Domb, 'Faith and Reason in Judaism', in Mott (ed.), *Can Scientists Believe?*: 131.

17. 'The Promise of Chaos': 15.

18. Jonah 1–4.

19–20. *Confessions*, Book XI, 'In the beginning God created . . .', trans. F. J. Sheed, in Cahn (ed.), *Classics of Western Philosophy*: 350.

21. Ibid., 360.

22. John 8:58.

23. Quoted in Dugald Murdoch, *Niels Bohr's Philosophy of Physics* (Cambridge: Cambridge University Press, 1987): 52.

24. p. 333.

25. P. Speciali (ed.), *Albert Einstein and Michele Besso, Correspondence 1903–1955*, letter of 12 December 1952 (Paris: Hermann, 1972): 453. Quoted in A. Pais, 'Subtle is the Lord . . .': *The Science and the Life of Albert Einstein* (Oxford: Clarendon Press, 1982): 382.

26. Quoted in E. L. Mascall, *Christian Theology and Natural Science* (London: Longman, 1956): 180.

27. Job 38:2.

28. Job 40: 2, 8.

29. Romans 9:20.
30. *Brideshead Revisited* (Penguin Books, 1986; first published by Chapman & Hall, 1945): 387.
31. Isaiah 55:8.
32. John 8:14.
33. 'What is Chaos?': 352.
34. I Corinthians 13:12.
35. 'The Invincible Ignorance of Science', presented as the Eddington Memorial Lecture at Cambridge in January 1988; first published in *Contemporary Physics*, Vol. 29, No. 4; reprinted in *Great Ideas Today: 1990* (London: Encyclopedia Britannica, 1990): 337.

CHAPTER 7

1. London: Dent, 1983: 6.
2. *One World*: 26.
3. *Confessions*, trans. R. S. Pine-Coffin (London: Penguin Books, 1961): 27 (Book 1, Chapter 6).
4. New York: Longmans, Green & Co., 1936.
5. *Reason and Reality* (London: SPCK, 1991): 5.
6. *This Hebrew Lord* (New York: Seabury Press, 1974): 14.
7. Deuteronomy 4:29.
8. This and the quoted passages that follow it are from C. S. Lewis, *The Lion, the Witch, and the Wardrobe*, Book 1 in *The Chronicles of Narnia* (New York: Macmillan Publishing Co., 1950): 45–7.
9. *An Enquiry Concerning Human Understanding*, in Cahn (ed.), *Classics of Western Philosophy*: 838.
10. Ibid., 840.
11. *Science and Providence*: 55.
12. *An Enquiry Concerning Human Understanding*: 841.
13. Ibid., 847.
14. 'The Invincible Ignorance of Science': 337.
15. *The Unauthorized Version: Truth and Fiction in the Bible* (London: Viking Penguin, 1991): 38.
16. Ibid., 7.
17. From a sermon delivered at St Peter's Church, Morristown, New Jersey, in April 1993.
18. *Science and Religion*: 8.
19. Ibid., 37.
20. *Proslogion*, trans. William E. Mann, in Cahn (ed.), *Classics of Western Philosophy*: 368.

21. This and the other quotations from Aquinas that follow it are from *Summa Theologica*, Part 1, Question 2, Third Article, in Cahn (ed.), *Classics of Western Philosophy*: 400–1.
22. Personal letter to the author.
23. *A Brief History of Time*: 174.
24. Aquinas, op. cit., Part 1, Question 1, Eighth Article: 392.
25. See Polkinghorne, *Reason and Reality*: 28.
26. Quoted in Brooke, *Science and Religion*: 208.
27. Tennessee Williams, *The Night of the Iguana* (New York: New Directions, 1961): 59.
28. 'What is Chaos?': 354.

CHAPTER 8

1. *A Brief History of Time*: 175.
2. Proverbs 2:1–6.
3. Personal letter to the author.
4. Quoted in M. Mitchell Waldrop, 'The Quantum Wave Function of the Universe', *Science*, 242, 2 December 1988: 1250.
5. *Theories of Everything*: 210.
6. 1 Corinthians 13:12.

BIBLIOGRAPHY

Abbott, Larry F. 'Baby Universes and Making the Cosmological Constant Zero', *Nature*, 336, 22 and 29 December 1988: 711–12.

Adler, Mortimer J. 'Reality and Appearances', from Mortimer Adler, *Ten Philosophical Mistakes*. London: Macmillan Publishing Company, 1985.

Alexander, P. 'Complementary Descriptions', *Mind*, Vol. LXV, 1976.

Anselm of Canterbury. *Proslogion*, trans. (1977) William E. Mann, reprinted in Steven M. Cahn (ed.), *Classics of Western Philosophy*. Cambridge: Hackett Publishing Co., 1977, 3rd edition, 1990.

Aquinas, Thomas. *Summa Theologica*. Questions 1 and 2 of Part 1 are reprinted in Steven M. Cahn (ed.), *Classics of Western Philosophy*. Cambridge: Hackett Publishing Co., 1977, 3rd edition, 1990. The translator's name is not given.

Augustine of Hippo. *Confessions*, trans. R. S. Pine-Coffin. London: Penguin Books, 1961.

Barrow, John D. *Pi in the Sky: Counting, Thinking, and Being*. Oxford: Clarendon Press, 1992.

Barrow, John D. *Theories of Everything: The Quest for Ultimate Explanation*. Oxford: Clarendon Press, 1991.

Barrow, John D., and Tipler, F. J. *The Anthropic Cosmological Principle*. Oxford: Oxford University Press, 1986.

Begley, Sharon. 'How is a Quark Like a Frisbee?', *Newsweek*, 19 April 1993: 53.

Begley, Sharon. 'Is Science Censored?', *Newsweek*, 28 September 1992.

Berry, R. J. (ed.). *Real Science, Real Faith*. Eastbourne: Monarch, 1991.

Brooke, John Hedley. *Science and Religion: Some Historical Perspectives*. Cambridge: Cambridge University Press, Cambridge History of Science Series, 1991.

Burrell, David B., and Bernard McGinn. *God and Creation: An Ecumenical Symposium*. Notre Dame, Indiana: University of Notre Dame Press, 1990.

Casti, John L. *Searching for Certainty: What Scientists Can Know About the Future*. New York: William Morrow and Co.

Chandrasekhar, Subrahmanyan. *Truth and Beauty: Aesthetics and Motivations in Science*. Chicago: University of Chicago Press, 1987.

Colodny, Robert G. (ed.). *Frontiers of Science and Philosophy*. Pittsburgh: University of Pittsburgh Press, 1962.

Coveney, Peter, and Highfield, Roger. *The Arrow of Time*. London: W. H. Allen, 1990.

Davies, Paul C. W. *God and the New Physics*. London: Dent, 1983.

Davies, Paul C.W. *The Mind of God: Science and the Search for Ultimate Meaning*. London: Simon & Schuster, 1992.

Dawkins, Richard. *The Blind Watchmaker: Why the evidence of evolution reveals a universe without design*. London: W. W. Norton & Co., 1986.

Dimopoulos, Savas, Raby, Stuart A., and Wilczek, Frank. 'Unification of Couplings', *Physics Today*, October 1991: 25–33.

Dirac, Paul. 'The Evolution of the Physicist's Picture of Nature', *Scientific American*, May 1963: 45–50.

Dodd, James E. *The Ideas of Particle Physics: An Introduction for Scientists*. Cambridge: Cambridge University Press, 1984; reprinted and revised 1988.

Eccles, J. C. *The Understanding of the Brain*. New York: McGraw-Hill, 1973.

Eddington, Arthur S. *The Expanding Universe*. New York: Macmillan Company, 1933.

Eddington, Arthur S. *The Nature of the Physical World*. Cambridge: Cambridge University Press, 1928.

Einstein, Albert. *Out of My Later Years*. New York: Carol, 1956, 1984.

Einstein, Albert. 'Physics and Reality', *Journal of the Franklin Institute*, 221, 1936.

Ferguson, Kitty. *Black Holes in Spacetime*. New York: Franklin Watts, 1991.

Ferguson, Kitty. *Stephen Hawking: Quest for a Theory of Everything*. London: Bantam Press, 1992.

Feynman, Richard. *QED: The Strange Theory of Light and Matter*. Princeton: Princeton University Press, 1985.

Ford, Joseph. 'A Complex World: Can We Cope?', in Georges and Pierre Lochak, *Courants, Amers, Écueils, en Microphysique*. Paris: Fondation Louis de Broglie, 1993.

Ford, Joseph. 'What is Chaos That We Should Be Mindful of It?', in Paul C. W. Davies (ed.), *The New Physics*. Cambridge: Cambridge University Press, 1989.

Fox, Robin Lane. *The Unauthorized Version: Truth and Fiction in the Bible*. London: Viking Penguin, 1991.

Freedman, David H. 'Maker of Worlds', *Discover*, July 1990: 46–52.

Galilei, Galileo. *Dialogues Concerning Two New Sciences* (1638). Macmillan edition, 1914.

Gingerich, Owen. 'Let There Be Light: Modern Cosmogony and Biblical Creation', in Roland Mushat Frye (ed.), *Is God a Creationist?* New York: Charles Scribner's Sons, 1983.

Gleick, James. *Chaos: Making a New Science*. London: Penguin Books, 1988.

Gribbin, John. *In Search of the Big Bang*. London: William Heinemann, 1986.

Guth, Alan, and Steinhardt, Paul. 'The Inflationary Universe', in Paul C. W. Davies (ed.), *The New Physics*. Cambridge: Cambridge University Press, 1989.

Hammond, Allen (ed.). *A Passion to Know: Twenty Profiles in Science*. New York: American Association for the Advancement of Science, 1984; New York: Scribner, 1985.

Hanson, N. Russell. *Perception and Discovery*. London: Freeman Cooper, 1969.

Hardy, Godfrey H. *A Mathematician's Apology*. Cambridge: Cambridge University Press, 1940.

Hartle, James B., and Hawking, Stephen W. 'The Wave Function of the Universe', *Physics Review*, D31, 1777.

Hawking, Stephen W. *A Brief History of Time: From the Big Bang to Black Holes*. London: Bantam Press, 1988.

Hawking, Stephen W. 'Baby Universes II', *Modern Physics Letters*, A, 5, 7, 1990: 453–66.

Hawking, Stephen W. 'Black Holes and Their Children, Baby Universes', Hitchcock Lecture, University of California, Berkeley, April 1988. Reprinted as 'Black holes and Baby Universes', in *Black Holes and Baby Universes, and Other Essays*. New York and London: Bantam, 1993.

Hawking, Stephen W. 'The Edge of Spacetime', in Paul C. W. Davies (ed.), *The New Physics*. Cambridge: Cambridge University Press, 1989.

Hawking, Stephen W. 'Einstein's Dream', lecture delivered at the Paradigm Session of the NTT Data Communications Systems Corporation in Tokyo in July 1991. Reprinted in Hawking, *Black Holes and Baby Universes*.

Hawking, Stephen W. 'Is Everything Determined?', lecture at the Sigma Club seminar at Cambridge, April 1990. Reprinted in Hawking, *Black Holes and Baby Universes*.

Hawking, Stephen W. 'My Position', a talk at Gonville and Caius College, Cambridge, in May 1992. Reprinted in Hawking, *Black Holes and Baby Universes*.

Hawking, Stephen W. 'The Origin of the Universe', lecture at the Three Hundred Years of Gravity conference in Cambridge, June 1987. Reprinted in Hawking, *Black Holes and Baby Universes*.

Hawking, Stephen W. 'Wormholes in Spacetime'. Unpublished, August 1987.

Hawking's 'History of Time' Re-considered. Symposium, The Cambridge Review, March 1992. Contributors: John Polkinghorne, Malcolm Longaire, Michael Redhead, Jonathan Powers, Stephen Hawking.

Heisenberg, Werner. *The Physicist's Conception of Nature*. London: Greenwood Press, 1958.

Hofstadter, Douglas R. *Gödel, Escher, Bach: An Eternal Golden Braid*. Hassocks, Sussex: Harvester Press, 1979.

Hume, David. *An Enquiry Concerning Human Understanding*, reprinted in Steven M. Cahn (ed.), *Classics of Western Philosophy*. Cambridge: Hackett Publising Co., 1977, 3rd edition, 1990.

Isham, Christopher J. 'Creation of the Universe as a Quantum Process', in Robert J. Russell, William R. Stoeger, S.J., and George V. Coyne, *Physics, Philosophy, and Theology: A Common Quest for Understanding*. Vatican City State: Vatican Observatory, 1988.

Jaki, S. *The Road of Science and the Ways to God*. Edinburgh: Scottish Academic Press, 1972.

Jastrow, Robert. *God and the Astronomers*. London: W. W. Norton & Co., 1978; reprinted 1992.

Kant, Immanuel. *Grounding for the Metaphysics of Morals*, trans. Paul Carus (1902), revised by James W. Ellington, reprinted in Steven M. Cahn (ed.), *Classics of Western Philosophy*. Cambridge: Hackett Publishing Co., 1977, 3rd edition, 1990.

Lederman, Leon M. *The God Particle: if the Universe is the Answer, What is the Question?* New York: Houghton Mifflin Company, 1993.

Lederman, Leon M., and Schramm, David N. *From Quarks to the Cosmos: Tools of Discovery*. New York: Scientific American Library, 1989.

Lemaître, Georges. *The Primeval Atom*, trans. Betty H. Korff and Serge A. Korff. Copyright 1950. Reprinted in Timothy Ferris, *The World Treasury of Physics, Astronomy, and Mathematics*. New York: Little, Brown and Co., 1991.

Lewis, C. S. *God in the Dock*, ed. Walter Hooper. Grand Rapids, Michigan: William B. Eerdmans Publishing Company, 1970.

Lewis. C. S. *The Lion, the Witch and the Wardrobe*, Book 1 in *The Chronicles of Narnia*. New York: Macmillan Publishing Co., 1950.

Lewis, C. S. *Miracles: A Preliminary Study*. New York: Macmillan Company, 1947.

Lewis, C. S. *Surprised by Joy: The Shape of My Early Life*. London: Harcourt, Brace and World, 1955.

Lightman, Alan. *A Modern Day Yankee in a Connecticut Court, and Other Essays on Science*. New York: Viking; Harmondsworth, Middlesex: Penguin, 1986.

Longaire, Malcolm. 'The New Astrophysics', in Paul C. W. Davies (ed.), *The New Physics*. Cambridge: Cambridge University Press, 1989.

Maxwell, James Clerk, 'Does the progress of Physical Science tend to give any advantage to the opinion of Necessity (or Determinism) over that of the Contingency of Events and the Freedom of the Will?', essay written in 1873, reprinted in Lewis Campbell and William Garnett, *The Life of James Clerk Maxwell*. London, 1882; New York: Johnson Reprint Corporation, 1969.

Mehra, Jagdish (ed.). *The Physicist's Conception of Nature*. Norwell, Mass.: Kluwer Academic Publishers, 1973.

Moore, James R. 'Charles Darwin Lies in Westminster Abbey', *Biological Journal of the Linnean Society*, 1982: 97–113.

Mott, Nevill (ed.). *Can Scientists Believe? Some Examples of the Attitude of Scientists to Religion*. London: James & James, 1991.

Murdoch, Dugald. *Niels Bohr's Philosophy of Physics*. Cambridge: Cambridge University Press, 1987.

Newman, John Henry. *The Idea of a University*. Reprinted by Doubleday & Co., New York, 1959.

Osler, Margaret J., and Farber, Paul Lawrence (eds.). *Religion, Science, and Worldview: Essays in Honor of Richard S. Westfall*. Cambridge: Cambridge University Press, 1985.

Page, Don N. 'Hawking's Timely Story', *Nature*, 332, 21 April 1988: 742–3.

Pais, A. *'Subtle is the Lord . . .': The Science and the Life of Albert Einstein*. Oxford: Clarendon Press, 1982.

Paley, William. *Natural Theology – or Evidences of the Existence and Attributes of the Deity Collected from the Appearances of Nature* (1802). Reprinted by St Thomas Press, Houston, Texas, 1972.

Paul, Iain. *Science and Theology in Einstein's Perspective*. Edinburgh: Scottish Academic Press, 1986.

Peacock, Roy E. *A Brief History of Eternity: A Considered Response to Stephen Hawking's 'A Brief History of Time'*. London: Monarch Publications, 1989.

Peitgen, H.-O., and Richter, P. H. *The Beauty of Fractals*. Berlin and Heidelberg: Springer-Verlag, 1986.

Peitgen, H.-O., and Saupe, D. *The Science of Fractal Images*. Berlin: Springer-Verlag, 1986.

Penrose, Roger. *The Emperor's New Mind: Concerning Computers, Minds, and the Laws of Physics*. Oxford: Oxford University Press, 1989.

Penrose, Roger. 'Gravitational Collapse and Space-Time Singularities', *Physics Review Letters*, 14, 57–59, 1965.

Pippard, Brian. 'Eddington's Two Tables', in *Great Ideas Today: 1990*. London: Encyclopedia Britannica, 1990: 311–17.

Pippard, Brian. 'The Invincible Ignorance of Science'. Presented as the Eddington Memorial Lecture at Cambridge in January 1988; first published in *Contemporary Physics*, Vol. 29, No. 4; reprinted in *Great Ideas Today: 1990* (London: Encyclopedia Britannica, 1990): 324–37.

Poincaré, Henri. *The Foundations of Science: Science and Hypothesis, The Value of Science, Science and Method*. New York: Science Press, 1929.

Polkinghorne, John. *One World: The Interaction of Science and Theology*. London: SPCK, 1986.

Polkinghorne, John. *Reason and Reality: The Relationship Between Science and Theology*. London: SPCK, 1991.

Polkinghorne, John. *Science and Creation: The Search for Understanding*. London: SPCK, 1988.

Polkinghorne, John. *Science and Providence: God's Interaction with the World*. London: SPCK, 1989.

Popper, Karl R. *The Logic of Scientific Discovery*. London: Hutchinson, 1961.

Prigogine, Ilya, and Stengers, Isobel. *Order out of Chaos*. London: Heinemann, 1985.

'The Promise of Chaos: An Interview with Georgia Tech Physics Professor Joseph Ford', *Georgia Tech Research Horizons*, Spring 1988: 10–15.

Ray, Christopher. *Time, Space, and Philosophy*. London: Routledge, 1991.

Sandage, Allan. 'Cosmology: The Quest to Understand the Creation and Expansion of the Universe', in Byron Preiss, *The Universe*. New York: Bantam Books, 1987.

Selvin, Paul. 'How Do Particles Put on Weight?', *Science*, Vol. 259, 8 January 1993: 173–4.

Silk, Joseph. *The Big Bang*. New York: W. H. Freeman & Co., 1989.

Smith, Robert W. *The Expanding Universe: Astronomy's 'Great Debate'*. Cambridge: Cambridge University Press, 1982.

Tierney, John. 'Subrahmanyan Chandrasekhar: Quest for Order', in Allen L. Hammond (ed.), *A Passion to Know: Twenty Profiles in Science*. New York: Charles Scribner's Sons, 1985.

Torrance, Thomas F. (ed.). *Belief in Science and in Christian Life: The Relevance of Michael Polanyi's Thought for Christian Faith and Life*. Edinburgh: Handsel Press, 1980.

Torrance, Thomas F. *Reality and Scientific Theology*. Edinburgh: Scottish Academic Press, 1985.

Trower, W. Peter. 'Muddling to Discovery', *Newsweek*, 24 August 1992: 52.

van den Beukel, A. *More Things in Heaven and Earth: God and the Scientists*, trans. John Bowden. London: SCM Press, 1991.

Vanauken, Sheldon. *A Severe Mercy*. London: Walker and Co., 1977.

Waldrop, M. Mitchell. 'The Quantum Wave Function of the Universe', *Science*, 242, 2 December 1988.

Weinberg, Steven. *Dreams of a Final Theory: The Search for the Fundamental Laws of Nature*. New York: Pantheon Books, 1992.

Weinberg, Steven. *The First Three Minutes: A Modern View of the Origin of the Universe*. London: André Deutsch, 1977; reprinted and revised 1988.

Wheeler, John A. *A Journey into Gravity and Spacetime*. New York: W. H. Freeman & Co., 1990.

Wilford, John Noble. 'Scientists Report Profound Insight on How Time Began'. *New York Times*, 24 April 1992: 1, 16.

Wilford, John Noble. 'Weighing Milky Way, Astronomers Find Halo of Dark Matter', *New York Times*, 8 June 1993: C1.

Ziman, John M. *Public Knowledge: An Essay Concerning the Social Dimension of Science*. Cambridge: Cambridge University Press, 1968.

INDEX

Absolute values, 80–81
Absurdity, 6
Abundance of elements in universe, 99–100
Aesthetics of science see Beauty
Albalag, Isaac, 267
Alien science, 79
Alpher, Ralph, 98–9
Alternate explanations, 84–6, 148–9, 151–2
American Astronomical Society, 91
Anarchy, as reality, 16
Andromeda nebula, galaxy, 64, 93
Animal population, study of fluctuations in, 210–12, 215
Anselm, St, of Canterbury, 266–7
Anthropic principle, 140, 164–6, 177–8, 223–5
and baby universes, 171–3
Antimatter, 18–19, 56, 103–4
Approximate descriptions, theories, 26, 32, 128, 132, 225, 237
Aquinas, St Thomas, 125, 131, 266–71

Argument from design, see Religious evidence
Articles of faith, 75, 97, 269
see also Assumptions underlying science and religion
Artificial intelligence, 181–2
Artificial life, 182
Aspect, Alain, 71–2
Assumptions, 7, 79
Friedmann's, 93–95
of our existence, 10, 11, 240
questioning of, 12–34, 65
underlying science and religion, 8–11, 37, 40, 74, 79, 113, 134, 240, 269–70
Asymmetry, 17–19
and origin of matter, 104
see also Symmetry
Atheism, 73, 86, 126
Atom, 13–14, 27, 46, 106–7
and uncertainty principle, 13–14, 27, 106–7
Attractors, 210–11
Lorenz attractor, 211

Attractors (*cont.*)
  strange, 211, 213
Augustine, St, of Hippo, 97, 140,
  225–6, 244–5
  on eternity, 140, 225
  and literal interpretation of
    Genesis, 261
  on time, 140, 225–6

Baby universes, 114–16, 127–9,
  139, 171–3
  and anthropic principle, 171–3
Bach, J. S., 238
Background radiation of the
  universe
    discovery of, 99
    prediction of, 98
Barrow, John D., 65, 129, 132,
  170, 235, 282
  on contradictions in
    mathematics, 65
  Pi in the Sky, 65
  Theories of Everything, 40, 170
Bayreuth Festival, 35–6
Beauty
  in physics, 33, 113, 172, 180,
    187
  in science, 59–63, 87, 133, 187,
    232
Belief in both God and science, 2,
  145, 184, 280–81
Bell, John, 72
Bell-Aspect experiment, 21, 72
Bell Laboratories, 98
Bergman, Ingmar, 201, 221
Bernstein, Leonard, 62, 85, 159
  Mass, 62
Beukel, A. van den, 72

Bible
  and argument from
    explanatory power, 270–71,
    273–4
  as evidence, 242, 260–63
Big Bang, 74–6, 96, 99–100,
  102–8, 126–7, 164
  and entropy, 119–22
  evidence supporting, 98–100,
    104–6, 126
  history leading to acceptance of
    theory of, 89–100
  and inflationary universe
    theory, 168–9
  opposition to, 74–6
  problems with, 103–6, 170
  relevance for religious belief,
    74–6, 96–8, 126–7, 203
  singularity, *see* Singularities
  theory and reality, 126
Big Crunch, 122
Black holes, 22–4, 101
  and Chandrasekhar and
    Eddington, 67–8
  definition of, 22
  mathematical beauty of, 61
  naming, 27, 101
  and singularities of infinite
    density, 22–4, 43, 101, 107,
    113
Bohm, David, 72
Bohr, Niels, 21, 222
  and wave-particle duality,
    228–31
Boltzmann, Ludwig, 133
Bondi, Hermann, 75, 96
Bosons, 46–7
Boundary conditions, 47–8, 136

Boundary conditions (*cont.*)
  in no-boundary proposal, 76,
    112–13, 126, 128, 137–9
  see also initial conditions
Brideshead Revisited, 232
Bridge, J. Frederick, 2
Brooke, John Hedley,
  on Bohr, 230–31
  on Darwin, 39
  on Galileo, 265
Burke, Bernard, 99
'Butterfly effect', 208

Casual arguments, 125, 267–70
Causality, 12–15
Cause and effect, 12–16, 19
Cepheids, 93
CERN, 52, 55, 56, 175
Challenger explosion, 12, 15
Chandrasekhar, Subrahmanyan,
  66–8, 101
Chaos theory, 19, 25, 32, 74, 162,
  201, 204–21, 235, 275–7
  definition of, 206
  and evolution, 275–6
  predictability in, *see*
    Predictability
  relevance for belief in God,
    215–18, 220–21, 275–7
  relevance for determinism and
    predictability, 25, 32, 78,
    213–19, 220–21, 235, 275–7
  and unpredictable systems, *see*
    Unpredictable systems
Christ, resurrection of, 194
Clockwork universe, 221, 275
*Close Encounters of the Third
  Kind*, 185

Coleman, Sidney, 172–3
Colossians, quoted, 139
Common sense, *see* Reality
Complementarity, 228–31,
  280–81
Complexity theory, 19, 25, 32, 74,
  78, 162, 201, 204–21, 224
  and chaos theory, 214
  relevance for determinism and
    predictability, 25, 32, 214
Computability, 132, 182
Computers
  human minds explained as,
    181–3
  program to demonstrate
    evolution, 152–8
  and randomness, 206–7
  simulation of fluctuations in
    animal populations, 210–12
  simulation of weather, 208, 211
Constants of nature, 24, 25, 48,
  80–81, 128, 173, 175–8
Contingency of universe, 8, 32,
  177, 190
  in chaos theory, 213–15, 276–6
Contraditions in mathematics, *see*
  Mathematics
Contradictions in nature, 33
Contrary truths, 26–7, 29
Copernicus, Nicholas, 163
Cosmic background explorer
  (COBE), satellite, 106
Cosmic background radiation, 99,
  105–6
Cosmological constant, 92
'Cosmological constant problem',
  171–3
Coveney, Peter, 213–14

Creation
  and the Big Bang, 74–5, 96–8,
    126–7
  chicken-and-egg dilemmas of,
    134–42, 159
  ex nihilo, 123–4
  'free lunch', 123–4
  Genesis account of, 194–5
  moment of, 12–13, 96–8
Creator, *see* God
Crucifixion, 83
Cubism, 58–9
cummings, e. e., 115
Cyclical universe, 122–4

Dark matter, 117–19
Darwin, Charles, 1–3, 38–9, 150,
    157, 277
  funeral, 1–3
  see also evolution
Darwin, Erasmus, 71
Davies, Paul C.W., 21, 72, 112,
    128, 133, 140, 150, 241
Dawkins, Richard
  and argument from design,
    150–51
  The Blind Watchmaker, 76
  challenges belief in a designer
    God, 76, 149–52, 176–8, 224
  computer program
    demonstrating evolution,
    152–5
  discussion of evolution, 149–63
  and emergence of self-
    awareness, 161
  and 'Godlessness of science',
    76
  on probability of life emerging,
    158, 276

  on probability of miracles,
    201–2
de Broglie, Louis, 61
de Sitter, Willem, 92–3
de Sua, F., 65
Death in Venice, 186
Deconstructionism, 262
Derby Philosophical Society, 71
Determinism, 30–33, 82, 200,
    215–16, 220–21
  on the quantum level, 200
  'top-down', 221
DeWitt, Bryce, 145
Dicke, Robert, 99
Dimopoulos, Savas, 174–6
Dirac, Paul, 59–61, 231
DNA, 71, 157–9, 162, 177, 276
Duhem, Pierre, 53

Eccles, John, 199
Ecclesiastes, quoted, 185
Eddington, Sir Arthur, 5, 10,
    66–8, 93, 94
  *The Nature of the Physical
    World*, 5
Edison, Thomas, 19–20
Einstein, Albert, 20–22, 143,
    258–9
  on concept of logical
    consistency, 62
  and the cosmological constant,
    92, 171
  and ether, 173
  and expanding universe, 75, 92,
    94
  formula $E = Mc^2$, 124
  on 'gift of fantasy', 38
  and quantum theory, 21–2, 61,
    67, 71–2, 222

Einstein, Albert (*cont.*)
    on religion, 143, 146
    theory of general relativity,
      92–4, 236;
      see also General relativity
    and warp of spacetime, 49,
      114, 141, 171
    and wave-particle duality,
      228–9, 231
Electromagnetic force, 47, 49,
    51–2, 171
Electromagnetic radiation, 16
Electrons, 13, 14, 42, 46–7, 174–6
    in early universe, 102, 174
    electric charge of, 164
    mass and charge, as a constant
      of nature, 24, 48, 164–5
    in uncertainty principle, 14, 107
    in weak force, 54
Electroweak theory, and
    verification of, 45–6, 49–56, 61,
    76, 175, 202
    and Higgs particle, 175
    influence on physics, 54
    neutral current in, 52–3
Eliot, T. S., 20, 57, 237, 277
Eliptical orbits of planets, 180
Emergence of life
    statistical probability of,
      158–60, 162
Energy conservation at origin of
    universe, 124
Energy in vacuum, 123–4, 171–4
Entropy, 119–22, 207
Equus, 186
Eternity, St Augustine's view,
    225
Ether, 173
Everett, Hugh, 222

Evil, problem of, 6, 179, 274
Evolution, 2, 16, 17, 29, 82,
    149–63, 175–7, 197
    and argument from design, 2,
      150, 155–6, 176–7
    and belief in God, 2, 149–50
    and chaos, 275–6
    computer program to
      demonstrate, 152–5, 160
    and creation of human beings,
      150–52, 177
    and emergence of self-
      awareness, 161, 176
    as evidence of a universe
      without a designer, 76,
      149–63, 176–7
    evidence supporting, 157
    and human mind, 16, 17, 60,
      82, 181–3
    and laws of physics, 156
    and possibility of prediction,
      82, 151–2, 155
    as source of aesthetics and
      morality, 60, 87
    vs. God as creator, 149–63,
      176–7

Faith in God, 7, 82, 86, 137, 151,
    187–9, 194, 205, 241, 247, 249,
    264, 266, 269, 271–2, 279, 281
    Bible and, 261
    consistent with faith in science
      or not, 1–2, 145, 184, 280–81
Faith in science, *see* Science
Falsifiability, *see* Science
Fanny and Alexander, 186
Fantasy, role in science, 38, 41–2
Feigenbaum, Mitchell, 211
Fermi, Enrico, 46

Fermilab, 52
Fermions, 46–7
Feynman, Richard, 30–31
First Cause, 63, 131, 135–7,
    143–5, 267, 158–9
  candidates, 141–2, 158–9, 233
  choice of First Cause candidate
    as act of faith, 136–7, 145
First principles, 269
Forces of nature, 46–7, 164
  unification of, 47–8; see also
    Electroweak theory, Weak
    force, Gravitational force,
    Electromagnetic force,
    Strong force
Ford, Joseph, 282
  and chaos, 206, 207, 209, 214,
    235, 275–6
  on evolution, 275–6
  on God as a Mississippi
    riverboat gambler, 217–18
Fox, Robin Lane, 261–2
Fractals, 211
Free will, 30–33, 181, 190,
    218–20, 226
Friedmann, Alexander, 92–5, 98
  models of the universe, 117–19

Galaxies, 18, 62, 89–90, 93, 95,
    99–102, 105, 109, 115
  clustering, 25
  and dark matter, 117–18
  recession, 23, 93–4
  in steady state theory, 96
Galaxy clusters, 25, 105
Galilei, Galileo, 64, 282
  and the Joshua miracle, 192–3
  on maths and nature, 64
  upholding science and God,
    265

Garnow, George, 98–9
Gell-Mann, Murray, 40, 60
  on beauty in physics, 60
General relativity, 4, 5, 23–4,
    91–4, 101, 102, 126, 134, 236
  and beauty, 60
  breakdown at singularities,
    23–4, 102
  distinction between space and
    time dimensions, 111
  prediction of expanding
    universe, 92, 100
  prediction of singularities, 22,
    24, 101–2, 106–7
  and quantum theory, 62, 107,
    123, 171–2
Genesis account of creation, 145,
    194–5
  interpretation in Judaeism, 195
Geometry in nature, 17–18
Glashow, Sheldon, 46
Gluons, 47
God
  and absolute values, 81
  as answer to the 'Why' of the
    universe, 84
  believer-dependent, 29
  Biblical, 185–240
  and Big Bang theory, 126–7
  and chaos and complexity
    theory, 215–21
  creation in image of, 8, 161
  as designer of human beings,
    149–63, 176–7
  as embodiment of laws of
    physics, 145–6
  evidence for and against, 241;
    see also Religious evidence
  and evolution, 149–62, 176–7

God (*cont.*)
faith in, 6–7, 145
falsifiability of, 44, 74, 179, 184
as Final Cause, 224
as fine-tuner of initial
  conditions, 163–78
as First Cause, 137, 139–45,
  158–9, 267
as fundamental rationality
  behind universe, 146–7
'hands-off' policy, 179, 218
and human choice, 218–20
intervening in the universe,
  79–90, 214, 217–20
as law-breaker, 189–90
and mathematical and logical
  consistency, 62, 64, 131
mind of, 3, 6, 8, 11, 19, 84, 87,
  90, 144, 146
and no-boundary proposal,
  134–42
proof of, 74, 267; *see also*
  Religious evidence
as purposeful creator, 146–9
relevance of science for belief,
  125–9
'rival good to', 232
as self-confirming hypothesis,
  231–4
as 'senile delinquent', 273
simplicity/complexity of, 62,
  85, 159, 224
as source of laws of physics,
  146–7
testimonies of, 248–51
within, and working through,
  human beings, 180–83
God-of-the-Gaps theology, 183,
  204–5

Gödel, Kurt, 64–6, 132
Gödel's incompleteness theorem,
  64–6, 132
Gold, Thomas, 75, 122
Gordian knot, 102, 106, 164–6,
  171
Gravitational force, 46–7, 164,
  168
Graviton, 46
Gravity, 49–50, 105; *see also*
  Gravitational force
  on the moon, 79
  as a repulsive force, 168–9
Guth, Alan, 123, 168; *see also*
  Inflationary universe
  and 'free lunch', 123

Hall, Lawrence, 174
Hanson, Russell, 44
Hardy, G. H.,
  on mathematical beauty, 60
  on mathematical reality, 63, 87
Hartle, Jim, 109, 111, 128–42
Hawking, Jane, 133
Hawking, Stephen W., 9, 38, 84,
  86, 122, 141
  agnosticism of, 26, 76, 109
  and anthropic principle, 166,
    170–73
  baby universes, 114–16, 127–9,
    139, 171–3
  and black holes, 24
  A Brief History of Time, 3, 6,
    30, 76, 108–9, 145
  concept of God, 145–6
  and determinism and
    predictability, 30–32
  and Mind of God, 3, 20, 76, 84,
    86, 147, 279

Hawking, Stephen W. (*cont.*)
  no-boundary proposal, 76,
    108–13,
      127–8, 134–42; *see also*
      Boundary conditions, Initial
      conditions, Imaginary time
  on objective reality, 26, 132
  and positivist position, 132
  and singularities, 101, 106–8
  Stephen Hawking: Quest for a
    Theory of Everything, 4
  and Theory of Everything, 20,
    30–32, 282
  triumph over adversity, 4, 264
  and wormhole theory, 114–16,
    127–9, 171–3
Haydn, F. J., 88
Heart of Darkness, 186
Heisenberg, Werner, 39, 228
Heisenberg uncertainty principle,
    13–15, 27, 61, 106–8, 111, 114,
    123, 171, 205
Herman, Ralph, 98–9
Higgs field, 174–6, 178
Higgs particle, 175–6
Highfield, Roger, 213–14
Hoffding, H., 231
Holmes, Sherlock, 193
Holy Ghost, 109
Hoyle, Fred, 75, 86, 96, 164
Hubble, Edwin, 23–4, 93–4
Human condition, 6
Human mind and personality
  scientific explanation of, 82–3,
    181–3
Human perception and
  consciousness, 16
Human soul, 83, 95
Human values, 239–40

Humason, Milton, 93
Hume, David, 254–9
Hyracotheria, 223–4
Imaginary horses, 35, 184, 283
Imaginary numbers, 109–10
Imaginary time, 111–15, 132–3,
    139
'Infinite Pass-the-Parcel', 21
Inflationary universe, 128,
    167–71, 178
  and anthropic principle, 170
  eternal version, 169
  and 'horizon problem', 170
  and Theory of Everything, 170
Initial conditions, 31, 48, 81, 108,
    112, 120, 128, 136–7, 139,
    208–9
  and the anthropic principle,
    164–70
  in chaos theory, 208–18
  fine-tuning of, 163–4
  and inflationary universe
    theory, 167–70
  see also Boundary conditions

James, William, 231
Jastrow, Robert, 74, 96–8
  on reaction among scientists to
    Big Bang, 74, 97
Job, 186, 232
Johnson, Samuel, 4
Jonah, 218–20
Joshua stops the sun, 184, 192–4

Kant, Immanuel, 231, 271–3
Kepler, Johannes, 178, 180
Khalatnikov, Isaac, 100
Kierkegaard, Soren, 231
Kuchar, Karel, 140–41

Kupfer, Harry, 35–6

La Grange, Joseph Louis, 277
La Place, Pierre Simon de,
    209–10, 213, 216, 277
Large Hadron Collider (LHC),
    175
Large Magellanic Cloud, 117–19
Laws of nature, 25, 33, 49, 195–6
    and God, 274–7
Leap of faith, 3
Ledermann, Leon, 145
    The God Particle, 145
Leibniz, Gottfried, 109, 232
Lemaître, Georges, 89, 92–4, 96,
    101
Leptons, 174–6
Levels of complexity, 82
Lewis, C. S., 86–8, 230, 249, 271
    on evaluating uncorroborated
        evidence, 253–4
    on resolving contradiction
        between Biblical and scien-
        tific views of creation, 230
Liddon, Canon H. P., sermon on
    Darwin, 1–2
Life, sacredness of, 83
Lifshitz, Evgenil, 100
Light
    speed of, in a vacuum, as a
        constant of nature, 24
    travel faster than, 22
    as waves or particles, 228–31
Lightman, Alan, 108–9
Lin, Douglas, 117
Linde, Andrei, 169
Logical consistency, *see*
    Mathematical and logical
    consistency

Lorenz, Edward, 209, 211–13
Lorenz attractor, 211
Lowell Observatory, 90
Lucas, George, 71

Maimonides, Moses, 267
Malaria, 70
Mandelbrot, Benoit, 211
Mandelbrot set, 211
Mass, origin of, 173–6
Mathematical and logical
    consistency, 44, 60–66, 133–4
    created by God?, 178
    as First Cause, 63, 131–7,
        141–5, 158, 177
    as stronger concept than God,
        62–4, 131
    as stronger evidence than
        experimental and
        observational evidence, 64
    see also Theory, scientific
Mathematical reality, 63–4,
    129–34
Mathematics
    beauty in, 61, 133
    contradictions in, 33, 65, 129
    correlation with nature, 64
    faith in, 62, 64–5, 243
    as a human invention, 130
    as independent reality, 130–31
    as inherent in nature, 130–31
    philosophies of, 129–34
    proof in, 63–5
    as a self-consistent system of
        logical deductions and
        connections, 131
    truth of, 133
    universality of, 64
    as working like a computer, 13

Matter
  and antimatter, 103–4
  distribution of, *see* Universe
  origin of, 103–4
  as particles or waves, 228–9
  see also Dark matter
Maxwell, James Clerk, 133, 209
May, Robert, 210–12, 214–15
McVittie, George, 93
Mehra, Jagdish, 59–60
Merrivale, Sir Henry, 85
Midrash, 215
Miller, John, 91
Mind, *see* Human mind
Mind of God, *see* God
Mind's-eye-view, *see* Reality
Miracles, 185–9, 231–2, 243, 250,
  259
  definitions, 191
  Hume and, 253, 255–9
  Joshua and the sun, 192–4
  and natural processes, 192
  Polkinghorne and, 199
  public and private significance
    of, 243
  scientific explanations for, 2,
    191, 193, 201–4
Monty Python, 77
  *Monty Python and the Holy
    Grail*, 35–6
Morgan, Jim, 30
Mott, Sir Nevill, 190
Mount Wilson Observatory, 93
Music, rules in, 237–8

Natural theology, 86, 88; *see also*
  Religious evidence
Neptune, discovery of, 42

Neutral current, in weak nuclear
  force, 52–3
Neutrons, 46, 174
Newman, Cardinal John Henry,
  151
Newton, Isaac, 30, 42, 78, 173,
  196, 205–6, 214, 235–6, 277
Newtonian dynamics, 205–6, 235
No-boundary proposal, *see*
  Hawking
Non-computable functions, 132

Objective reality, *see* Reality
Objective truth, *see* Reality

Page, Don
  on need for creator in no-
    boundary proposal, 139–40
Paley, William, 150, 163, 277
Paul, St, quoted, 139–40, 237,
  282–3
Pauli, Wolfgang, 228
Peieris, Rudolph, 72
Penrose, Roger, 100–2, 121–2
  and black holes, 24, 100, 101
  The Emperor's New Mind, 182
  and entropy in universe, 120–22
  on evolution of consciousness,
    161–2
  and expanding universe, 106
  against possibility of explaining
    the mind as a super-complex
    computer, 182
  on quantum world, 200
  and singularities, 101
Penzias, Arno, 98–9, 106
Perspective, 18
Photon, 46, 49, 51–2, 98, 171–2

Pippard, Sir Brian, 10, 11, 239, 277
  'The Invincible Ignorance of Science', 259–60, 277
  on public and private knowledge, 259–60
  on scientific explanation of the human mind, 182
Planck, Max, 2
Poincaré, Henri, 35–6, 38, 59, 112
Point of view in science, *see* Science
Political correctness, 70–71
Polkinghorne, Revd Dr John, 134, 147
  on Hume, 257
  on no-boundary proposal, 134
  on quantum theory and God, 135–6, 199, 200, 243, 247, 266, 268, 278
Popper, Karl, 43, 75
Powers, Jonathan, 64, 66
Prayer, 181
  for healing, 73–4
  see also Miracles
Predictability, 8, 24, 31, 34, 78, 82, 188, 199–200, 202–7, 226, 276
  and chaos and complexity theory, 25, 32, 205–21
  and evolution, 82, 151–2, 155
  as a goal of scientific discovery, 26, 78
  loss of at singularity, 23–4, 102
  of orbits in solar system, 25, 235, 275
  on quantum level, 13–14, 21, 27, 124, 199–200

of religious experience, 243
Predictable systems, 40, 78, 205; *see also* Unpredictable systems
Prigogine, Ilya, 213
Primeval atom hypothesis, 92–3
Proof in mathematics, *see* Mathematics
Proof in science, *see* Science
Protons, 46–7, 174
Protostars, 90
Proverbs, quoted, 2, 279
Psalms, quoted, 179–80
Pulsing universe, 117–22

Quantum energy fluctuation, 114, 123–4, 171
Quantum gravity, 107–13, 123
Quantum state, 199–200
Quantum theory, 4, 13, 21–2, 27–9, 71–2, 123, 227
  breakdown of predictability in, 82
  and general relativity, 62, 107, 123, 171–2
  many-worlds view of reality, 222
  see also Heisenberg uncertainty principle
  see also Reality and quantum theory
Quarks, 46, 174–6
Quasars, 95, 100

Raby, Stuart, 174–6
Radiation, electromagnetic, 16, 98–9
Randomness, 206–7, 216
  and God's foreknowledge, 216

Reactions to suggestion that God intervenes in the universe, 187–9

Reality
  'as-it-is-in-itself', 11, 79–81, 129
  and common sense, 4, 5, 9, 16, 29, 71–2, 129–30
  conflict of realities, 228–31
  differing views of, 78–9
  direct encounter with, 38–9, 64
  Hawking on, 132
  independent, 38–9, 41, 222
  limited by what science can study, 84, 132–3
  many-world's view of, 222
  mathematical, *see* Mathematical reality
  mind's-eye view of, 6, 10, 79–80, 280
  objective, 25–9
  observer-dependent, 26–9
  and quantum theory, 5, 10, 39, 71–2, 108, 222, 223
  representations of, 57–9
  scientific theories and, 125–9

Red shift as evidence of recession of galaxies, 90–94
Reductionism, 221
Relativity, *see* General relativity
Religious evidence, 87–8, 143–5, 184, 189, 229, 230, 241–78, 282
  argument from availability, 277–8
  argument from design, 2, 150–52, 159–66, 176–8
  argument from explanatory power, 270–73
  argument from nature, 274–7
  argument from reason, 266–70

Biblical, 261–3
  experiential, 246–51, 270, 282
  of individual witness, 247–60, 282
  ontological proofs, 267
  as 'private knowledge', 242–7
  of results in individual lives and society, 263–5

Responding to God, 161–2, 178
Resurrection of Christ, 194
Richards, Paul, 99
Royal Society of England, 8
Rubbia, Carlo, 55–6

Sacredness of life, 83
Salam, Abdus, 46, 49–53
Salam-Weinberg theory, *see* Electroweak theory
Salvation, 83
Sandage, Allan, 75
Saul, Handel, 2
Schrödinger, Erwin, 61
Schrödinger's equation, 199–200, 202
Science
  assumptions underlying, *see* assumptions underlying science and religion
  basic principles of, 38
  beauty in, *see* Beauty
  creativity in, 38, 56
  and direct experience of the universe, 38, 56, 77
  dogma-of-the-leading-edge of, 36
  elite of, 66–9
  faith in, 35, 36–7, 71, 75, 97, 82, 134, 136, 145, 187, 194, 281

falsifiability in, *see* Theory,
  scientific
Godlessness of, 73–4, 76
logical consistency in, 41
moral content of, 69–70, 87–8,
  144
point of view in, 39–41, 44–5,
  52–9, 66–77, 78–9
proof in, 26, 37, 43, 75, 85
as public knowledge, 38, 242
revealing relationships, 59, 80
right to be wrong, 37
role of experiment and
  observation in, 41, 52–3
simplicity in, 33, 61–2; *see also*
  Theory, scientific
social and cultural influence in,
  69–72
spirit of the times in, 69–71
theory in, *see* Theory, scientific
'verdict of', 36
Scientific explanation of human
  mind and consciousness, 82,
  181–3
Scientific method, 8, 25–6, 34,
  36–44
  evolution of, 79
  limits of, 78–85
Second Law of Thermodynamics,
  119–22, 207
Self-awareness, emergence of, 161
Sensitive dependence on initial
  conditions, 201, 208–19
  in quantum theory, 201
Seurat, 45, 57–8
Seuss, Dr, 90
The Seventh Seal, 221
Singularities, 22, 196
  at beginning of universe, 23,

100–2, 112
  at Big Bang, 76, 100–2
  breakdown of physical laws at,
    33, 74–5, 102
  failure of prediction at, 24
  prediction of in general
    relativity, *see* General
    relativity
  and quantum theory, 106–8
  as slammed door, 22, 100–2
  undermining of, 103, 106, 134
  see also Black holes
Slipher, Vesto, 90–93
Smoot, George, 105–6
Solar system
  orbits, 13, 180
  prediction of orbits, 25
Soul, 82–3
Spacetime
  origin, 124–5
  warping of, 22, 141, 171–2
Spectacles-behind-the-eyes, 44–5,
  52–9, 246–7
Speed of light, 95
  travel, 22
  in a vacuum, as a constant of
    nature, 24
Spielberg, Steven, 71, 185–6
Spiral nebulae, 90–91, 93–4
Spong, Bishop John S., 249, 263
St John's College, Cambridge,
  59–60
Steady state theory, 75, 96, 98,
  100
Stengers, Isobel, 213
Stokes, Sir George, on miracles,
  231–2
String theory, *see*
  Supersymmetric string theory

Strong force, 46–7
Supernatural, 83, 85
Superstrings, 61, 119, 128; *see also* Supersymmetric string theory
Supersymmetric string theory, 42
Symmetry, 33, 176
  in nature and human design, 17–19
Symmetry-breaking, 49–52, 202–4, 227

Taylor, John, 72
Theory, scientific
  criteria for judging worth of, 40–44, 59
  development of, 40–43
  falsifiability of, 42–4, 74–6, 129, 135, 137
  mathematical and logical consistency in, 40–44, 60–66, 128–9
  metaphysical, 43
  parsimony in, 42, 61
  and prediction, 43
  role of, 40–44
  sources of, 41–4
Theory of Everything, 20, 30–33, 45–9, 62, 81, 128, 282
  defined, 45–9
  and predictability, 81–2
Theory of relativity, *see* General relativity
Thomas, St ('Doubting Thomas'), 2
Time
  and St Augustine, 140, 225–6
  and black holes, 24
  as broken symmetry, 51

chronological flow of, 17, 120
  created by God, 226–7
  running backwards, 100
  as space dimension, 111
  see also Imaginary time
'Tough love', 6
Top quark, 175–6
Trower, W. Peter, 77
Tryon, Edward, 123
'Tyranny of old men', 68

Ultimate truth, 26, 37, 62, 66, 78, 81–2, 87, 131, 233
Uncertainty principle, *see* Heisenberg uncertainty principle
Universe
  accessibility to human minds, 8, 9, 19–25, 74–6, 134
  background radiation of, 98–9
  Big Bang model of, 102; *see also* Big Bang
  contingency of, 8, 30–33
  contracting, 119
  distribution of matter in, 18, 25, 105–6
  early history in Big Bang theory, 102–3, 105–6
  early twentieth-century view of, 90
  end of, 6; *see also* Big Crunch
  expansion of, 23–4, 74–5, 89, 90–95, 106
  as First Cause, 137–45
  as 'free lunch', 123, 127–9
  'horizon problem' in, 170
  inflation theory of, 167–71
  late twentieth-century view of, 89

observer-dependent, 28–9
origin of, 6, 12–13, 23, 31–2,
    74–6, 89, 92, 95–101, 115,
    123, 145
preconceptions about, 66
pulsing model of, 117–22,
    127–9
as 'put-up job', 163–4, 178
rationality of, 8, 9, 12–19
right- and left-handedness of,
    18
smoothness problem of, 104–6
unified description of, 33–4,
    134
uniformity problem, 106
unity of, 9, 33–4, 134
Unpredictable systems, 205, 274

Vacuum, 171–4
van der Meer, Simon, 55–6
Van Quine, W., 53
'Verdict of science', 26

W and Z particles, 55–6, 175
Wagner, Richard, 35–6
    The Ring of the Nibelungs, 35

Warp of spacetime, 49, 114, 141,
    171–2
Waugh, Evelyn, 232
Wave function, 199
Wave-particle duality, 228–31
Weak force, 46, 49–53
Weinberg, Steven, 46, 49–56, 74,
    77
    on beauty in physics, 60
Wheeler, John A., 27–9, 62, 72,
    105, 114, 117, 166
    on beauty in physics, 60
    and black holes, 101
    poem about mass and
        spacetime, 141
    Twenty Questions, 27–8
Wilbur, Richard, 4
Williams, Tennessee, on God,
    273
Wilson, Robert, 99, 106
Wobbling of nothingness, 123–5
World-view of Western science,
    7–9
Wormholes, 61, 114–15, 125,
    127–9, 139, 172–3, 178